AUTHENTIC OPPORTUNITIES FOR WRITING ABOUT MATH in High School

Teach students to write about math so they can improve their conceptual understanding in authentic ways. This resource offers hands-on strategies you can use to help students in grades 9–12 discuss and articulate mathematical ideas, use correct vocabulary, and compose mathematical arguments.

Part One discusses the importance of emphasizing language to make students' thinking visible and to sharpen communication skills, while attending to precision. Part Two provides a plethora of writing prompts and activities: Visual Prompts; Compare and Contrast; The Answer Is; Topical Questions; Writing About; Journal Prompts; Poetry/Prose; Cubing and Think Dots; RAFT; Question Quilt; and Always, Sometimes, and Never. Each activity is accompanied by a clear overview plus a variety of examples. Part Three offers a crosswalk of writing strategies and math topics to help you plan, as well as a sample anchor task and lesson plan to demonstrate how the strategies can be integrated.

Throughout each section, you'll also find Blackline Masters that can be downloaded for classroom use. With this book's engaging, standards-based activities, you'll have your high school students communicating like fluent mathematicians in no time!

Tammy L. Jones has taught students from first grade through college. Currently, she is consulting with individual school districts in training mathematics teachers on effective techniques for being successful in the mathematics classroom, supporting mathematics instruction, and STEM integrations. She is co-author of two book series published with Routledge: *Strategies for Common Core Mathematics* and *Strategic Journeys for Building Logical Reasoning*.

Leslie A. Texas has over 20 years of experience working with K–12 teachers and schools across the country to enhance rigorous and relevant instruction. She believes that improving student outcomes depends on comprehensive approaches to teaching and learning. She is co-author of two book series published with Routledge: *Strategies for Common Core Mathematics* and *Strategic Journeys for Building Logical Reasoning*.

Also Available from Tammy L. Jones and Leslie A. Texas
(www.routledge.com/k-12)

**Strategic Journeys for Building Logical Reasoning, K–5:
Activities Across the Content Areas**

**Strategic Journeys for Building Logical Reasoning, 6–8:
Activities Across the Content Areas**

**Strategic Journeys for Building Logical Reasoning, 9–12:
Activities Across the Content Areas**

**Strategies for Common Core Mathematics:
Implementing the Standards for Mathematical Practice, K–5**

**Strategies for Common Core Mathematics:
Implementing the Standards for Mathematical Practice, 6–8**

**Strategies for Common Core Mathematics:
Implementing the Standards for Mathematical Practice, 9–12**

AUTHENTIC OPPORTUNITIES FOR WRITING ABOUT MATH in High School

Prompts and Examples for Building Understanding

Tammy L. Jones and Leslie A. Texas

NEW YORK AND LONDON

Cover images: @ Getty Images

First published 2025
by Routledge
605 Third Avenue, New York, NY 10158

and by Routledge
4 Park Square, Milton Park, Abingdon, Oxon, OX14 4RN

Routledge is an imprint of the Taylor & Francis Group, an informa business

© 2025 Tammy L. Jones and Leslie A. Texas

The right of Tammy L. Jones and Leslie A. Texas to be identified as authors of this work has been asserted in accordance with sections 77 and 78 of the Copyright, Designs and Patents Act 1988.

All rights reserved. No part of this book may be reprinted or reproduced or utilised in any form or by any electronic, mechanical, or other means, now known or hereafter invented, including photocopying and recording, or in any information storage or retrieval system, without permission in writing from the publishers.

Trademark notice: Product or corporate names may be trademarks or registered trademarks, and are used only for identification and explanation without intent to infringe.

Library of Congress Cataloging-in-Publication Data
Names: Jones, Tammy L., author. | Texas, Leslie A., author.
Title: Authentic opportunities for writing about math in high school: prompts and examples for building understanding / Tammy L. Jones and Leslie A. Texas.
Description: New York, NY: Routledge, 2025. | Includes bibliographical references.
Identifiers: LCCN 2024022211 (print) | LCCN 2024022212 (ebook) |
ISBN 9781032449326 (hbk) | ISBN 9781032447865 (pbk) | ISBN 9781003374596 (ebk)
Subjects: LCSH: Mathematics—Authorship. | Mathematics—Study and
teaching (Secondary) Classification: LCC QA20.M38 J66 2025 (print) |
LCC QA20.M38 (ebook) | DDC 510.71/2—dc23/eng20240723
LC record available at https://lccn.loc.gov/2024022211
LC ebook record available at https://lccn.loc.gov/2024022212

ISBN: 978-1-032-44932-6 (hbk)
ISBN: 978-1-032-44786-5 (pbk)
ISBN: 978-1-003-37459-6 (ebk)

DOI: 10.4324/9781003374596

Typeset in Warnock Pro
by codeMantra

Access the Support Material:
https://resourcecentre.routledge.com/books/9781032449326 or visit https://resourcecentre.routledge.com and search for the book's ISBN, title or authors.

Special thanks to Pixaby, WordArt, Padowan Graph, PRESENTERMEDIA, Pexels, and Geometer's Sketchpad for certain images used in this book.

We would like to give a special thanks to Trevor Styer for his work to ensure the graphics used throughout the series were high quality and reproducible for classroom use.

Online Resources

Several of the resources in this book are available online as free downloads so you can print them for classroom use. To access them, find the book at the url below and search for this book's ISBN, title, or authors. Note that you will be asked to provide information from the book before you can obtain the downloads. https://resourcecentre.routledge.com/

You can also follow this direct link: https://resourcecentre.routledge.com/books/9781032449326

Contents

Meet the Authors — xi

Preface: A Note to Our Readers — xiii

PART ONE
Why Writing in Math Matters — 1

Chapter 1   Purposeful Writing: Intentional Design — 3

PART TWO
Writing Prompts — 15

Chapter 2   Visual Prompts — 17

Chapter 3   Compare and Contrast — 27

Chapter 4   The Answer Is... — 33

Chapter 5   Topical Questions — 39

Chapter 6   Writing About... — 57

Chapter 7   Journal Prompts — 112

Chapter 8   Poetry/Prose — 118

Chapter 9   Cubing and Think Dots — 121

Chapter 10 RAFT — 139

Chapter 11 Question Quilt — 144

Chapter 12 Always, Sometimes, and Never — 147

Contents

PART THREE
Planning and Implementation — **167**

Chapter 13   Crosswalk — 169

Chapter 14   Bringing It All Together — 171

Afterword — 183

Bibliography — 185

Meet the Authors

Collectively, Tammy and Leslie have almost 45 years of classroom experience teaching in elementary, middle, high school, and college. This has included urban, suburban, rural, and private school settings. Being active members of their professional organizations has allowed them to continually grow professionally and model lifelong learning for both their students and their peers. In their 30-plus years of combined consulting work, they have had opportunities to work with teachers and students from kindergarten through college level. This work has spanned almost all 50 states. Their work has included helping to develop standards and curriculum at the state level as well as implementing curriculum and best practice strategies at the classroom level. One of the things that sets Tammy and Leslie apart as consultants is their work with classroom teachers, modeling and offering continued support throughout the year to build capacity at the building and district levels. Tammy and Leslie co-authored the 2013 series from Eye On Education/Routledge-Taylor & Francis Group, *Strategies for Common Core Standards for Mathematics: Implementing the Standards for Mathematical Practice* (Grades K–5, 6–8, and 9–12), and the 2017 series from Routledge-Taylor & Francis Group, *Strategic Journeys for Building Logical Reasoning: Activities Across the Content Areas* (Grades K–5, 6–8, and 9–12).

An educator since 1979, **Tammy L. Jones** has worked with students from first grade through college. Currently, Tammy is consulting with individual school districts in training teachers on strategies for making content accessible to all learners. Writing integrations as well as literacy connections are foundational in everything Tammy does. Tammy also works with teachers on effective techniques for being successful in the classroom. As a classroom teacher, Tammy's goal was that all students understand and appreciate the content they were studying; that they could read it, write it, explore it, and communicate it with confidence; and that they would be able to use the content as they need to in their lives. She believes that logical reasoning, followed by a well-reasoned presentation of results, is central to the process of learning, and that this learning happens most effectively in a cooperative, student-centered classroom. Tammy believes that learning is experiential and in her current consulting work creates and shares engaging and effective educational experiences.

Leslie A. Texas has over 25 years of experience working with K–12 teachers and schools across the country to enhance rigorous and relevant instruction. She believes that improving student outcomes depends on comprehensive approaches to teaching and learning. She taught middle and high school mathematics and science and has strong content expertise in both areas. Through her advanced degree studies, she honed her skills in content and program development and student-centered instruction. Using a combination of direct instruction, modeling, and problem-solving activities rooted in practical application, Leslie helps teachers become more effective classroom leaders and peer coaches.

Preface

A Note to Our Readers

Our previous two book series, Strategies for Common Core Mathematics: Implementing the Standards for Mathematical Practice (Grades K–5, 6–8, and 9–12) and Strategic Journeys for Building Logical Reasoning: Activities Across the Content Areas (Grades K–5, 6–8, 9–12), provided a set of strategies and sample tasks that teachers could implement across the curriculum to engage students at a deeper cognitive level required by the rigorous college and career ready standards.

When we took on writing this new series, we asked ourselves: What is it that teachers want and would support students in becoming better communicators of mathematics? During training with teachers on our other two series, we often were asked how teachers could get more classroom-ready materials, such as questions, writing prompts, etc., that would support their work with students on writing and reading mathematics. Therefore, we wanted to create a collection of items for educators that would be practical and versatile, easy to implement, and yield results.

For the student, we created a collection of visual prompts that provide opportunities to engage in mathematics through looking at pictures of and from the world. There is an assortment of examples supporting the academic vocabulary associated with each math topic. Also included are ready-to-use writing prompts covering a variety of topics across the grade band. Sets of non-typical questions are provided to promote developing a deeper understanding of mathematics. Examples of various writing styles, including creative writing, meet the needs and interests of a diverse classroom.

For educators, it is important to understand students can only become comfortable (and proficient) communicating about mathematics by practicing it regularly. Today's high-stakes assessments require students to understand mathematics in context and to explain their reasoning behind strategies and solutions. There are enough strategies included to incorporate often (daily/weekly). Using these prompts and tasks is easy once the teacher has determined the instructional goals and targeted standards for implementation. There is teacher autonomy in implementing, but the prompts and tasks are ready to be used immediately.

These are great strategies for providing a variety of ways to engage students in mathematical discourse. The materials are versatile in use as handouts, visual displays, gallery walks, electronic documents, etc. The Crosswalk shows examples by mathematical topic. Teachers will find strategies for authentically integrating different writing techniques in the mathematics classroom, including creative writing. A sample lesson incorporating a number of these prompts and examples is included along with unique strategies and examples for differentiation in the mathematics classroom.

PART ONE

Why Writing in Math Matters

CHAPTER 1

Purposeful Writing: Intentional Design

Communication is essential in expressing ideas clearly and effectively. Language serves as a framework for that communication. Mathematics is often said to have its own language using symbols in addition to words. Combining mathematical language with written/spoken language can often provide deeper insight into how information is being processed, connections that are being made, conclusions drawn, etc. This data is important in assessing understanding as well as moving thinking further.

This book will look at how writing can be used in the following:

1. Making student thinking visible – formative assessment
2. Building communication skills while attending to precision – construct a viable argument and critique the reasoning of others (Standard of Mathematical Practice 3) and attend to precision (Standard of Mathematical Practice 6)
3. Establishing authentic reasons for writing, not just so we can say we did write in math.

As introduced in our book series *Strategic Journeys for Building Logical Reasoning: Activities Across the Content Areas* (Jones & Texas, 2017), there are seven opportunities for writing. These served as a guideline and informed the choices made regarding the types of writing included in this series.

Authentic Opportunities for Writing about Math in High School

- **Making Meaning** – understanding the question posed and identifying given and needed information necessary to proceed
- **Showing Evidence** – using facts and/or data to support one's argument/hypothesis/work
- **Reflecting** – being metacognitive with respect to strategies and/or processes
- **Inquiry** – creating questions to drive investigation and/or research
- **Educating** – informing others in various forms/purposes – persuasive, descriptive, expository, and narrative
- **Creating Ideas** – brainstorming/free writing to begin framing ideas
- **Producing Products** – using products to convey a message depending on audience and purpose (research papers, proposals, brochures, essays, public service announcements, etc.)

Making Student Thinking Visible: Formative Assessment

For teachers to elicit evidence of student understanding and provide feedback that moves the learning forward, students must be able to make their thinking visible. Many students struggle to organize their thoughts and capture their thinking on paper. Starting with a straightforward tool such as the Think-Write-Pair-Share (Jones & Texas, 2017) allows student specific guidance on where to begin writing. A blank piece of paper can mean "I don't know" or "I don't care." It is an important distinction, and providing tools for students to support the "I don't know" is critical in building their capacity to help themselves. In addition, this intentional emphasis on writing highlights the importance of being a good partner by bringing something to the table when coming together to discuss ideas. Below is one example of capturing this thinking.

Think, WRITE, Pair, Share

<u>Think</u> about…

<u>WRITE</u> about what questions come to mind in the area below.

<u>PAIR</u> with your partner and discuss what each of you wrote.

Be prepared to <u>SHARE</u> with the whole group.

Using with a Rich Task

The following tool can be used to help students organize their thoughts around a rich task. The task can be embedded so students can stay focused while engaging in the process. The "write" component gives very specific guidance to support students whether they understand the problem or not. It also provides stems for students to consider any time they are engaged in solving a problem.

Think-Write-Pair-Share	
Think Think about the problem. INSERT TASK HERE	**Write** Write by doing one of the following: If you can solve, choose a strategy, and solve. If you cannot solve… ❏ Write all facts you know about the problem ❏ Write anything you know related to the concept addressed in the problem ❏ Write questions you have about the problem
Workspace	
Pair Pair with a partner and take turns discussing your strategies and solutions. Use this space to record strategies that were different from yours.	**Share** Share various strategies and solutions with the group. Use this space to record strategies that were different from those of you and your partner.

Building Communication Skills While Attending to Precision

The following problem-solving process and graphic organizer were introduced in *Strategies for Common Core Mathematics: Implementing the Standards for Mathematical Practice* (Texas & Jones, 2013) and can Can be used to assist students in making sense of problems **(Standard of Mathematical Practice #1)** as well as decontextualizing and contextualizing word problems **(Standard of Mathematical Practice #2)**. The process also requires students to construct viable arguments **(Standard of Mathematical Practice #3)** as they formulate their own ideas about the meaning of the problem and make predictions about the outcome. Once a solution is obtained, students compare to the prediction to determine the reasonableness of the solution. By giving students explicit steps to unpack the problem, they begin the process with minimal to no teacher guidance and complete the initial steps. This eliminates the blank piece of paper or the famous, "I don't know" answer. Using a consistent process over time with students will assist them in becoming better problem solvers. While this process may not always "fit" every problem, it does help students develop a systematic approach to finding the "entry point" into various tasks.

The process is like the Three Reads strategy in that it asks students to read the problem more than once. The first time they read it in its entirety to understand the context of the problem. Steps 1 and 2 then ask students to reread specific sentences as they decode the text and make sense of the problem. See below for an explanation and how the graphic organizer is used to capture the process.

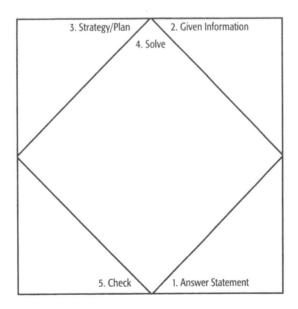

This organizer can be enlarged and copied onto paper for students. It can also be created by folding a piece of paper at each end as if making a paper airplane. Once opened, it will be partitioned as above.

For an interdisciplinary version, see our book *Strategic Journeys for Building Logical Reasoning: Activities Across the Content Areas* (Jones & Texas, 2017).

1. **Answer Statement**
    a. The question usually appears as the last sentence of the problem. Students can cover the other information and focus on the last line to determine what the problem is asking. (If the question is not here, students can check each preceding line until it is found.)
    b. Students write the question as an answer statement and leave a blank for the solution. Translating from a question to an answer statement can be challenging for some students. Practicing verbally asking and answering the question can assist in this process.
    c. Remind students to include the appropriate units for the context of the problem.
    d. The answer statement is a critical component and should be practiced even when not using this organizer. It ensures students understand the question being asked as well as guarantees they will answer the question posed if the problem contains multiple steps. Developing this habit promotes its transfer to testing situations and is particularly important when answering constructed response questions.

2. **Given Information**
    a. Students use the same process of viewing each sentence separately, covering everything else.
    b. Students determine and record relevant information from the problem.

3. **Strategy/Plan**
    a. Students use this space to state additional ideas they have about the problem, such as other information they know about the problem, possible strategies for getting started, estimations for the solution, constraints, or predictions.
    b. This is the section that allows students to formulate their own ideas about the problem and provides a place for them to create their own meaning about what is being asked.

c. Determining an estimate also provides a context for checking for reasonableness of the solution.
d. This step also allows students to become strategic problem solvers rather than impulsive ones by requiring them to consider the various strategies available and then determine which might be the most efficient to use in the given situation.
e. Many students are not versatile in the various problem-solving strategies available. Creating a Strategy Wall can be useful to build the students toolkit. See p. 28 for more information on Strategy Walls.

4. **Solve**
   a. Students select a strategy (translate verbal statements into mathematical statements, draw a picture, make a table, etc.) and solve.
   b. Students can compare their solution to the estimation to determine the reasonableness of their answer.

5. **Check**
   a. Students check their answers by substitution or by using another method to justify.
   b. This is also a good time to strategically partner students who used different strategies. Students can coach each other in the use of their strategy.
   c. Once the answer has been checked, students write the answer in the blank from Step 1.

Emphasis on Process over Solution

The purpose of any problem-solving process is to encourage students to think about the problem before impulsively jumping ahead to solving. It also encourages them to read and understand before assuming what is expected. To reinforce this point, students can be given a set of problems in which they are asked to complete the initial steps but not to solve. This allows the focus to be on making sense of the problem and planning before executing. If on a teaching team, this assignment can be completed in the ELA classroom since it involves decoding text and pre-writing skills. Once students have completed these initial steps, take away the problem set and have students complete the process by solving, checking, and answering the question. By not having access to the original problems, this will serve as an assessment of the initial steps. If students can complete the work, then the information gathered is sufficient. If not, it reveals key components that were overlooked.

Update and New Information

Since the publication of *Strategies for Common Core Mathematics: Implementing the Standards for Mathematical Practice* (Texas & Jones, 2013), many teachers have asked why the graphic organizer begins at the bottom right rather than the top left. There are two reasons it is organized in this manner. The first was in response to how the brain works when asked to attend. To focus and not just mindlessly record answers in a familiar sequence/order, the brain must consciously engage with the organizer and therefore students are more intentional with the process. The second reason was addressing when the process was internalized and the tool no longer needed. Most mathematics problems begin to be solved at the top left of the problem and then worked down to the bottom right where the solution usually is completed. This organizer begins with the end in mind (bottom right) and then comes full circle with the final answer.

The problem-solving process and the graphic organizer can be adapted to meet the needs of teachers and students and even eliminated as an organizer for students who internalize the process and no longer need the scaffold. Below is an example of a graphic organizer that was modified from the original. The first table contains scaffolds where there is a list of possible concepts/strategies for students to select as they build their toolkit. NOTE: The choices given here are general for illustration purposes and would be intentionally crafted for the specific unit in which it was being utilized. The second table has the supports removed.

Purposeful Writing: Intentional Design

Problem-Solving Process (Scaffolded)

The problem is asking me to…	I know…
Answer statement:	
Topic/concept this is related to… Ex. Combining like terms Solving system of equations Unit rates Proportional relationships	**Strategy for solving…** Draw a picture Guess and check Work backwards Use the standard algorithm Etc…
Solve (show work here)	
This solution means…	

Problem-Solving Process

The problem is asking me to...	I know...
Answer statement:	
Topic/concept this is related to...	**Strategy for solving...**

Solve (show work here)

This solution means...

**See Section "Problem-Solving Process" for Sample Lesson Plan using this tool

Purposeful Writing: Intentional Design

Questioning: A Tool for Promoting Communication

As discussed in section "Problem-Solving Process (Scaffolded)" of our second series, "Strategic Journeys for Building Logical Reasoning," there are opportunities for questioning students while working through the problem-solving process.

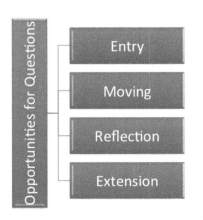

Entry: questions for students having difficulty getting started (Steps 1 and 2)
Moving: questions for places where students could get stuck (Steps 2–4)
Reflection: questions for students to use for metacognition after completing the problem/issue (Steps 4 and 5)
Extension: questions for students to engage in higher-order thinking skills with respect to the same concept and/or problem (after completing Step 5 and returning to Step 1)

These opportunities allow teachers to develop task-specific questions that can be used to support students as they are working through the process. In using these with students, it was noted that the first two opportunities occurred while students were in the middle of the process and the last two were once the process had been completed. Therefore, rather than viewing as four opportunities, they were condensed into two – I'm stuck, and I'm done.

I'm Done I'm Stuck

<u>Four Question Types</u>

1. Entry Questions
2. Moving Questions

3. Reflection Questions
4. Extension Questions

Task-specific questions can be generated and provided to students as needed, or they can be taught to access them on their own. If stuck, they can retrieve the appropriate questions that will allow them to move forward. For students who finish early, the done questions will be used to have them go deeper with the task rather than be assigned additional work, which is oftentimes seen as busy work.
See Section "Problem-Solving Process" for examples.

Establishing Authentic Reasons for Writing

Incorporating literacy across the curriculum has long been an emphasis in mathematical classrooms. Initially, this involved having students put into words how they solved the problem alongside the mathematical steps. Fortunately, the redundancy of this request was soon realized. The mathematics itself clearly articulated what students did to solve the problem. Therefore, students were asked to write about their thinking rather than what they did. For example, if solving an equation, students were asked to write about the properties of equality used to explain why they did rather than what they did.

Section "Building Communication Skills While Attending to Precision" introduces ten different strategies that can be used to provide authentic opportunities for students to write about math. Explanations of the strategies as well as content-specific examples have been provided to make these ready-to-use in the classroom. In addition, each provides the opportunity for differentiation. Below is a brief description of each:

Visual Prompts: pictures and images to initiate thoughts and discussions
Compare and Contrast: academic vocabulary word pairs to deepen understanding
The Answer Is…: giving an answer for which there could be multiple questions posed
Topical Questions: set of questions whose stems promote mathematical discourse
Writing About: using word clouds with academic vocabulary to write about a specific topic
Journal Prompts: assortment of ideas to engage students in journaling about mathematics
Poetry/Prose: collection of ideas to engage linguistic learners in expressing mathematical thought
Cubing and Think Dots: activities for independent learning
RAFT: creative writing opportunity
Question Quilt: an alternative way to present questions and provide student agency
Always, Sometimes, and Never: alternative way to view statements that promote critical thinking

PART TWO

Writing Prompts

CHAPTER 2

Visual Prompts

The visual prompts given here are actual photographs taken by the authors. These are different from what is known as "visual mathematics" which usually references the various visual representations in mathematics. The picture prompts harken back to *Mister Rogers' Neighborhood* Picture segments. These pictures help our students see the mathematics that is all around them. They also offer opportunities for students to engage in authentic mathematical communication.

The following collection of photographs can be used as journal prompts, discussion starters, bell ringers, or for centers, small groups, or learning stations. These pictures provide opportunities for students to engage in mathematics through looking at pictures of and from the world. As a starting point, have students free write what they see and describe it. This could be facilitated much like the *Notice and Wonder* prompts that the National Council of Teachers of Mathematics has brought to the forefront in the past few years.

High school students can name the geometric shapes they see used, the types of numbers they see, the type of function that might mirror parts of the pictures, and the mathematical topic that the image might conjure up.

Having students free to write about the visual prompts is ideal. You can make this a timed writing assignment where students must put writing implement to paper for a set amount of time, say 70 seconds. Beginning with a smaller amount of time and increasing it over the semester or year will help

Authentic Opportunities for Writing about Math in High School

students build stamina in free writing about a visual prompt. However, if some students need extra support, you can provide one of the following prompts:

- ❏ What do you see?
- ❏ How do you think math was used in this picture?
- ❏ What questions does the picture make you think about?
- ❏ What mathematical vocabulary could you use to describe the picture?
- ❏ Do you see any patterns in the picture? If so, describe the pattern.
- ❏ Where might you have seen something similar to what this picture is showing?
- ❏ What equations of lines would match the supports in the television tower?
- ❏ What functions would match the graphs of the wall mounting of interlocking circles?
- ❏ Discuss the path of the water shooting from the mouths of the frogs at the Dallas Arboretum?
- ❏ Research the works of Mexican artist Yvonne Domenge. How do you think she used mathematics when designing and making the Tabachin Ribbon?
- ❏ Describe geometrically the Amber Fort located in Jaipur, India.
- ❏ Describe geometrically the wall of windows on the side of a building in NY City.
- ❏ Mathematics is often called the tool of the sciences. Study the science wall found in a high school. Choose at least three of the science topics represented. What mathematics might be used in each of the topics you choose? Be specific.

Take your own pictures of things in your town or school. Look for open-source images on the internet. Have your students share photos. Remember that visual prompts offer all students a voice and provide an opportunity for most students to enter the conversation and make mathematical connections.

Visual Prompts

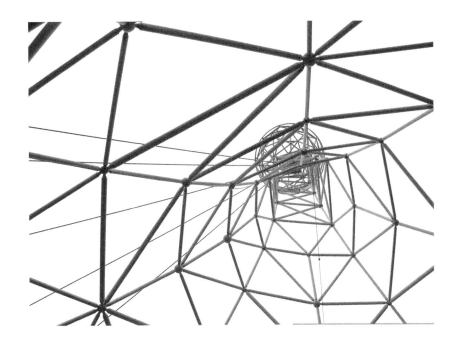

19

Visual Prompts

21

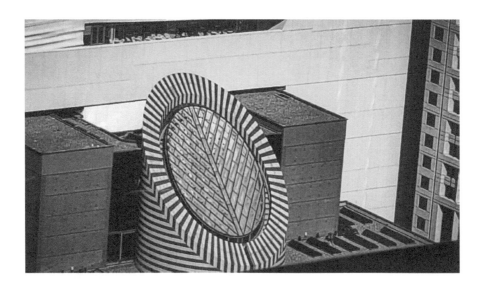

Visual Prompts

23

Visual Prompts

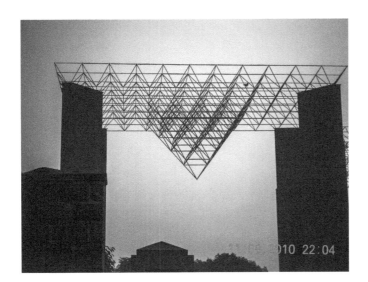

25

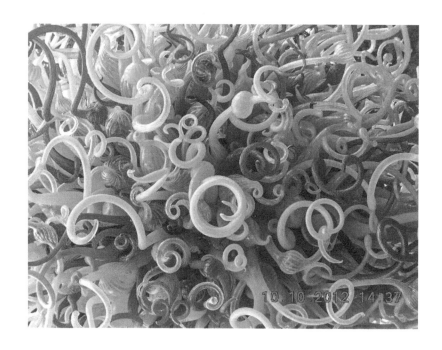

CHAPTER 3

Compare and Contrast

Writing math is typically a challenge for students. As discussed in our second series, "Strategic Journeys for Building Logical Reasoning," using the Mathematician's Notebook "can change the way you teach as well as how your students learn and experience their content. "The notebook becomes a dynamic place where language, data, and logical reasoning experiences operate jointly to form meaning for the student" (Jones). A Mathematician's Notebook helps students create an organized space for demonstrating their learning process. The notebook serves as a formative instructional tool as well as a portfolio of the students' learning experiences" (Jones & Texas, p. 14). Whether you are using a Mathematician's Notebook, an interactive notebook, or some other method of students chronicling their journey, all students need to be writing about math daily using paper and a writing implement.

Two of the main components of the Mathematician's Notebook are the glossary and the journal. Vocabulary is one of the foundations for developing an understanding of any subject area and mathematics is no exception. Students need many opportunities to use their vocabulary in their daily work. Having students develop a glossary and reference the glossary as they progress through the year provides a resource for the students to use in their current mathematics course as well as future courses. Additional opportunities for students to engage with their academic vocabulary are vital for students to develop the deep understanding needed for success.

One such opportunity is the Compare and Contrast activity. Students can simply make a T-chart on their paper. They write the word pair (or three columns if using three words), one word at the top of each column. Students then compare the words by listing the ways they are alike and different. They write their ideas in the columns below each word pair. They conclude by writing a summary sentence about their ideas. If time, students can complete additional pairs. There is a graphic organizer provided if desired to use. It is set up so when copied it can be cut in half and used with two students.

Interactive Word Walls and Strategy Walls

Ideally, the vocabulary used in this activity would already be displayed on a word wall of key terms that have been discussed throughout the instructional unit. A strategic way to make a word wall more interactive would be to use words from the wall for this activity. Assign students the words or allow student choice, which would reveal how students are making sense of the relationships between the concepts. Once the activity is complete, have students display their work on the wall alongside the words.

To reinforce the idea of students building a toolkit of strategies that can be used when problem solving, a strategy wall is a helpful anchor chart. Using words from the additional lists below (create a list, create a table, and create a graph, draw a picture and draw a diagram, educated guess and random guess, eliminate possibilities and solve a simpler problem, formula and function, look for a pattern and use a formula, work backwards and work forwards, write an equation or inequality and model with manipulatives), build a strategy wall at the conclusion of the activity by displaying the words (problem-solving strategies) and student responses.

A Beginning List of Word Pairs

Topic 1: Number and Quantity

Additive inverses and zero pairs
Arithmetic and algebraic
Cartesian plane and Argand plane
Directly proportional and indirectly proportional
Distance on the number line and distance between two points
Exponent and logarithm
Imaginary numbers and complex numbers
Magnitude of a vector and direction of a vector
Matrices and vectors
Matrix and determinant

Positive, negative, and zero (this is intentionally vague as it could related to integers, slope, the number line, etc.)
Radical and exponent
Rational numbers and irrational numbers
Real numbers and complex numbers
Rectangular form and polar form
Scalar quantity and vector quantity
Variable and coefficient
Variable and quantity
Whole number exponent and rational exponent

Topic 2: Algebraic Reasoning

Algebraic representation, tabular/numeric representation, graphical representation, and verbal representation of a function
Amplitude, period, and phase shift for trigonometric functions
Binomial Theorem and Pascal's triangle
Closed system and not closed system
Commutative Property and Associative Property and Distributive Property (Commutative and Associative can be for either addition or multiplication)
Composition of functions and inverse functions
Continuous and discrete
Continuous function and piece-wise function
Dependent variable and independent variable
Domain and range
Equations and functions
Equations, inequalities, and systems of equations
Evaluate and solve
Explicit function and recursive function
Exponential functions and logarithmic functions
Expressions and equations
Geometric sequence and arithmetic sequence
Graphs and tables
Horizontal asymptote and vertical asymptote
Identity and inverse
Linear functions and quadratic functions
Odd functions and even functions
Output and input
Polynomial functions and rational functions
Quadratic functions, absolute value functions, and polynomial functions

Reflexive property of equality, symmetric property of equality, and transitive property of equality
Relative maximum/relative minimum and absolute maximum/absolute minimum
Solution and answer
Solving a system of equations and solving a system of inequalities
Trigonometric functions and transcendental functions
Vertex form of a quadratic function and standard form of a quadratic function
Vertical shift of a function and the horizontal shift of a function
x-, y-intercepts and zeros of a function
Zeros of polynomial functions and factors of polynomials functions

Topic 3: Geometric Reasoning/Measurement and Units

Alternate interior angles along a transversal and alternate exterior angles along a transversal
Conditional statement, contrapositive, negation, and biconditional
Congruence and similarity
Congruent triangles and similar triangles
Converse and conditional
Corresponding angles and vertical angles
Draw and sketch and construct
Inductive reasoning and deductive reasoning
Inscribed circle and circumscribed circle
Linear and quadratic and cubic functions (use this opportunity to connect geometric reasoning to algebraic reasoning)
Orthocenter, centroid, circumcenter, incenter, and Euler's line
Parabola, circle, ellipse, and hyperbola
Parallel lines and perpendicular lines and skew lines
Rigid motion and isometry
Rotation, reflection, translation and dilation
Scale factor and dilation
Segment of a circle and sector of a circle
Supplementary angles and complementary angles
Vertical angles and adjacent angles

Topic 4: Data Analysis, Probability and Statistics

Addition rule for probability and multiplication rule for probability
Causation and correlation
Center and spread and shape
Compound events and simple events

Compare and Contrast

Conditional probability and basic probability
Expected value and mean
Frequency and relative frequency
Independent events and dependent events
Inference and observation
Linear association and nonlinear association
Making inferences and justifying conclusions
Measures of center and measures of variability
Positive association and negative association
Probability and odds
Single variable data and bivariate data
Two-way tables and sample space
Variance and standard deviation

Authentic Opportunities for Writing about Math in High School

Compare & Contrast

Choose a word pair. Write each word pair in the boxes below. Compare the words by listing the ways they are alike and different. Write your ideas in the columns below each word pair. Write a summary sentence about your ideas.

Word pair	
Compare:	
Contrast:	

Summary sentence(s):

Compare & Contrast

Choose a word pair. Write each word pair in the boxes below. Compare the words by listing the ways they are alike and different. Write your ideas in the columns below each word pair. Write a summary sentence about your ideas.

Word pair	
Compare:	
Contrast:	

Summary sentence(s):

CHAPTER 4

The Answer Is...

Students benefit from open-ended questions where there is possibly more than one correct response. This writing strategy allows students the opportunity to think beyond just procedural solving to get "the" answer. In some cases, the context is set up and given for the students. These questions will offer you as well as your students' insight into how they think about mathematics. Open-ended questions also encourage a growth mindset.

Students choose a card from "The Answer Is..." set to write about. Or you can assign one based upon students' individual needs. Students read the setup, if one is provided, then, they create a contextual problem for which the solution would be the answer given. This writing activity can be easily differentiated by setting parameters for students. The contextual problem can be single step, or it may be multi-step. It could require a specific operation or include quantities within specific parameters. Students could also be required to provide at least two different possibilities for a context where the solution is the answer. Drawings, illustrations, and labels might also be needed for a complete response.

Some examples of differentiation from the set provided.

❏ In Topic 1: Number and Quantity, students are given an answer dealing with interest; they can create scenarios that address simple interest and/or compound interest.

Authentic Opportunities for Writing about Math in High School

- ❏ In Topic 4: Data Analysis, Probability, and Statistics, when given three positive integers, students can choose to create a question looking for three specific numbers or a possible set of three numbers.
- ❏ In Topic 4: Data Analysis, Probability, and Statistics, the word that was used for the permutation was osmosis. This question was intentionally open-ended so other options of words can be used. Changing the word to, say, Mississippi, would return an answer of 34,650. Answers and words can be chosen based upon student needs.

The Answer Is...

Topic 1: Number and Quantity	
You are working with temperature changes. The answer is $x \leq -5°$.  What could the question be?	When dealing with interest, the answer you get is 30 years. 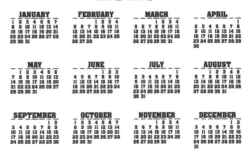 What could the question be?
In the study of electrical circuits, the formula $Z = \dfrac{V}{I}$ is used. It relates impedance (Z), voltage (V), and current (I). These can all be represented by complex numbers. The answer is $I = \dfrac{1}{2} + \dfrac{1}{4}i$. 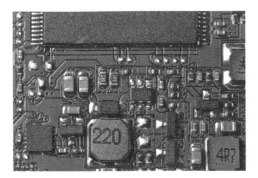 What could the question be?	The answer is … Volume = $5.7x^{\frac{1}{3}}$ cubic inches. What could the question be?

35

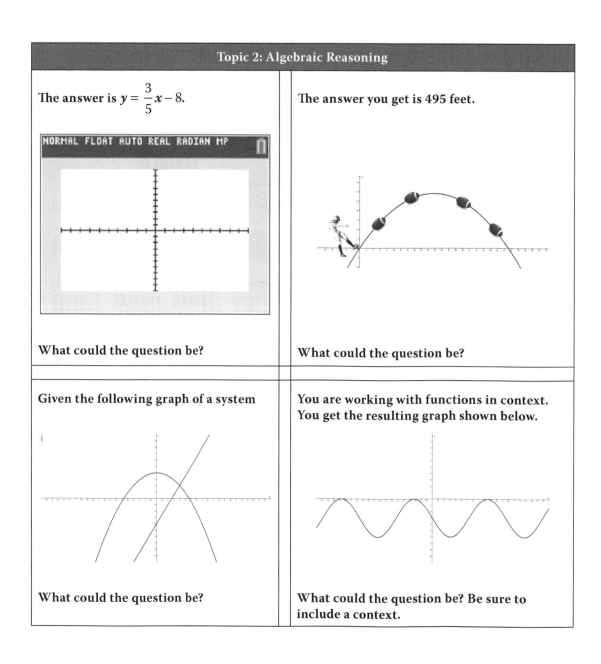

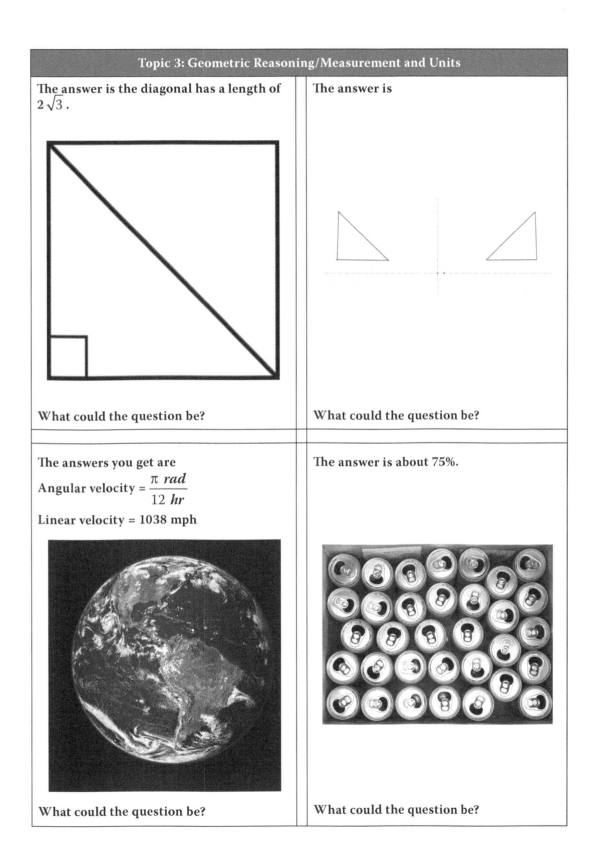

Topic 4: Data Analysis, Probability, and Statistics	
There are three positive integers. Their product is 288. The answer is that the mean is greater than 8. What could the question be?	Given a standard deck of cards, the answer is $P(B\|A) = \dfrac{1}{2}$. What could the question be?
For a given seven-letter word, the number of distinct ways in which the letters can be arranged is... 420. What could the word be?	In a section on shape, center, and spread of data, the graphical answer is... NORMAL FLOAT AUTO REAL RADIAN MP DRAW VERTICAL LINE X=6 Y=0.098337 STYLE What could the question be?

CHAPTER 5

Topical Questions

Questions are tools in teachers' toolbox and should be used as chisels to promote student thinking rather than pliers for answer-getting. Teachers practice, refine, and hone their questioning skills as they engage with students daily. Students can only provide a depth of answer based upon the quality of the question(s) asked.

As discussed in the section "Topic 3: Geometric Reasoning/Measurement and Units" of our second series, "Strategic Journeys for Building Logical Reasoning," there are opportunities for questioning students that naturally exist when students are working through a task or activity. And the questions need to be carefully crafted so the mathematical discourse is not shut down. Using a question stem such as "is," "do,", or "could" allows students the opportunity to simply answer yes or no and then they are done. If, however, you use a stem such as "how," "what," "when," or "where," along with others listed on our Q-Pyramid overlay (Jones & Texas, p. 92), you have opened the conversation and students must engage in mathematical discourse. This use of a more inquiring form of response, encourages students to justify or explain their responses, whether they be correct or incorrect.

As you hone your questioning technique, be aware that one of the most important parts of questioning is how you respond to students. You should respond to your students in a manner that supports their thinking as it reveals to you what and how they are thinking. Wait time is vital as a quick response can often shut down the individual's or rest of the class's thinking and/or reflection on what is being said. Asking students to explain why, or to further discuss how they thought about something may at first be a struggle with students, but if it becomes a consistent part of your questioning, students will eventually accept it.

The questions provided in this section are not universal, but rather nuanced to the topic they reference. By design, they require communicating the answer more fully and are perfect for encouraging students to write about math.

> NOTE: Don't limit yourself. Consider editing questions to include all parent functions, not just the ones included within the questions stems.
>
> Parent Functions: Linear, Quadratic, Cubic, Absolute Value, Reciprocal, Exponential, Logarithmic, Square Root, Sine, Cosine, Tangent

Topic 1: Number and Quantity

General

1. What is a number system?
2. What does it mean when a number system is closed?
3. How can a nested Venn diagram be used to represent the Real Number System?
4. How do irrational numbers relate to integers?
5. How can you determine if a decimal cannot be written as a rational number?
6. Why is pi considered an irrational number?
7. What is the difference in using an approximation for pi when calculating the area of a circle from using the exact irrational number?
8. David says that all numbers under a radical sign are irrational. Claudine disagrees. Who is correct and why?
9. How can the remainder theorem be used as a shortcut in polynomial division?
10. When the remainder is zero, what does it mean about the relationship between the dividend and divisor?
11. What role do remainders of polynomial division play in the factor theorem?
12. How can radicals help determine a length or distance on the coordinate plane?
13. How are these two expressions different: $(x+y)^n$ and $x^n + y^n$?
14. What types of numbers (e.g., integers, whole numbers, fractions, etc.) do you think indices of radicals could be? Why?
15. How is evaluating expressions with the natural base e similar to evaluating other exponential functions? How is it different?
16. When should you include an absolute value sign when simplifying a radical expression?
17. How can real number operations be extended to radical expression and equations?
18. How is adding two radicals similar to adding two variables? How is it different?
19. Does it matter whether you multiply radicals or simplify them first? Explain.
20. Could you ever multiply or divide radicals that have two different indices? Explain why or why not.
21. What exactly does rationalizing a denominator do to an irrational number in the denominator?

22. Under what conditions might you want to rationalize a denominator?
23. How could you determine if two radical expressions are like terms or not?
24. How can you represent numbers less than one using exponents?
25. How are the properties of exponents used to simplify and evaluate expressions, including expressions with rational exponents?
26. How can the properties of exponents be extended to rational exponents?
27. How can a nested Venn diagram be used to represent the Complex Number System?
28. How are complex numbers related to imaginary numbers and real numbers?
29. How can one use the patterns from powers of i to evaluate i^{123}?
30. How are commutative and associative properties applied when operating with complex numbers?
31. Why do complex roots always occur in conjugate pairs?
32. Which axis on the Cartesian plane is used for graphing the real part of a complex number?
33. Which axis on the Cartesian plane is used for graphing the imaginary part of a complex number?
34. How does the graphical representation for addition validates the numerical process for adding two complex numbers?
35. How are complex numbers used when working with Ohm's law?
36. How does the complex plane differ from the Cartesian plane?
37. What is an Argand diagram, and for what is it used?
38. How is multiplication with two complex numbers similar to binomial multiplication with real numbers? Different?
39. How can you explain the power property of logarithms and use it to solve equations?

Subtopic Specific

❑ **Vectors**

1. What two dimensions define a vector?
2. Under what conditions can two vectors be declared equal?
3. How can you show that vector addition is commutative?
4. How can you show that vector addition is associative?
5. Why is multiplication of a vector by a scalar another way of showing the distributive property for arithmetic?
6. What is the effect of scalar multiplication of a vector by a value of −1?
7. What is the difference between a vector and a scalar?

8. How do you think you would subtract vectors based on the work that you have done with vector addition?
9. How are the norm of a vector and the magnitude of a vector related?

❏ **Matrices**

1. What is a matrix?
2. How are matrices represented? In other words, what symbols would you use when writing a matrix?
3. What is the plural for the term matrix?
4. What is meant by the dimension of a matrix?
5. What is an entry in a matrix?
6. How do you add and subtract matrices?
7. When are matrix addition and matrix subtraction defined? Undefined? Why?
8. What is a zero matrix and how does it behave?
9. How do you multiply a matrix by a scalar?
10. What properties of real numbers can be extended to matrices?
11. How do the commutative and associative properties apply to matrices?
12. How do you think the distributive property applies to matrices?
13. What is the relationship between operations with matrices and operations with real numbers?

Topic 2: Algebraic Reasoning

General

- What are some ways to express a numerical relation?
- What would happen if a function machine provided two different outputs for one input?
- Could there be more than one way to write a function rule for a given input and output? Explain.
- What are the benefits of graphing functions on your calculator vs. by hand?
- Why would you use multiple representations of linear equations and inequalities?
- When do evaluate versus when do you solve when working with algebraic reasoning?
- What is a parent function and how does knowing the parent function help you describe transformations on the function?
- How can expressions be extended to create equations and/or functions? Give a specific example.
- How does the Cartesian coordinate plane help us visualize functions?
- What does it mean to have an equivalent expression?
- How can you determine if an algebraic algorithm is both effective and efficient?
- What is a variable?
- How are the trace feature and the table feature on the graphing calculators alike? How are they different?
- How do asymptotes relate to end behavior?

Subtopic Specific

- **Algebraic Functions** (Note: The first few questions have blanks that can be filled in with any desired function.)

    1. How does the shape of a/an _____ function relate to its domain? Its range?
    2. What is an example of a/an _____ function for which the domain and range are both $-\infty$ to $+\infty$?
    3. What is an example of a/an _____ function for which the domain is $2 < x < 3$?
    4. What is an example of a/an _____ function for which the range is $y = -52$?

Topical Questions

5. How can you identify key features of the graph of a/an _____ function from its equation?
6. Why is the shape of a/an _____ function different than the shape of a/an _____ function?
7. How could you describe the end behavior for a/an _____ function?
8. How many possible zeros would a/an _____ function have?
9. How can you use graphs of _____ functions to predict information about the future?
10. What is the difference between a function and an equation?
11. How can you tell if a graph can be classified as a function?
12. What is the difference between a function's dependent and independent variables?
13. What are some of the different notations that can be used to describe a function's domain including all numbers between 0 and 10?
14. What are the benefits of using various representations for a function? Be specific.
15. What does "=" mean in $f(x) = y$? Be very specific in your response.
16. What can you tell about a function by determining its average rate of change?
17. What is the difference between a relation (that is not a function) and a function?
18. Can a function's input value share an output value with a different input value? Explain.
19. How are patterns of change related to the behavior of functions?
20. How can patterns, relations, and functions be used as tools to best describe and help explain real life situations?
21. What does it mean to be a solution to an inequality?
22. What does it mean "to be true" when discussing the/a solution(s) to an equation? An inequality?
23. How can constraints or conditions in a real-world context be represented by inequalities? Explain using a specific example.
24. Which properties of equality do not hold for inequalities? Why? Show specific examples.
25. How do you represent absolute value functions?
26. Why are inequalities used when writing some absolute value models?
27. What does the solution to an absolute value mean?
28. What evidence is needed to determine how many solutions exist for an equation? For a system of equations?

29. What are the benefits and limitations of solving a system of equations using graphing, substitution, and elimination? Give an example of a system that would best be solved by each of the methods.
30. What are the identifying characteristics of a matrix in RREF?
31. When solving a 3x3 system, why would it be important to have a column in which the only non-zero entry is a "1"?
32. Can a 1x1 matrix be in RREF? Explain.
33. How do you interpret the output from the calculator when solving a 3x3 system?
34. What are some of the operational steps that can be taken to reduce a matrix to RREF by hand?
35. What clues can you look for within a word problem to set up a three-variable three-equation system?
36. What is the process going from a word problem to an augmented matrix to a solution?
37. If using a matrix when solving a system, how would you know if there was no solution for a given contextual problem?
38. What would a 3x3 system with infinite solutions look like?
39. A polynomial function of degree n can have up to how many turning points?
40. How does the graph of a root of odd multiplicity behave versus the graph of a root of even multiplicity in reference to the x-axis?
41. What is meant by the fluid nature of a polynomial model?
42. Why can't $a = 0$ for a radical function?
43. Why is it important to know the domain of a radical function?
44. What is the difference between an odd or even index of a radical?
45. What causes an extraneous solution to be introduced when solving a rational equation?
46. How can you tell if a solution is a valid solution or an extraneous solution?
47. How can you tell that a rational expression will have a slant asymptote?
48. What would be the identity function g for $(f+g)(x)=(g+f)(x)=f(x)$?
49. What is the difference between the horizontal line test and the vertical line test?

❏ **Transcendental Functions**

1. How are algebraic and transcendental functions different?
2. What does "transcendental" in transcendental function refer to?

Topical Questions

3. How could you determine the initial value of an exponential function? What does it signify physically?
4. How does the rate of growth of exponential functions differ from that of linear functions?
5. Under what conditions is it valid to set exponentiated terms equal to each other to form an exponential equation?
6. Describe the rate of growth of exponential functions. How does that differ from that of linear functions?
7. What characterizes exponential growth and decay?
8. How does the frequency of compounding affect compound interest? Give an example.
9. Why is the y-intercept so important when using exponential functions?
10. How can you explain a logarithmic scale being used to graph the Richter scale?
11. How can you explain a logarithmic function being used to model temperature change over time?
12. Keeping in mind what you know about logarithms, how do you explain the fact that Newton's law of cooling (or warming!) can be modeled logarithmically?
13. How do you determine the y-intercept of a logarithmic function?
14. How can you determine the asymptotes of logarithmic functions?
15. How can you use the fact that exponents and logs are inverses of each other to solve certain types of equations?
16. Why are some trigonometric functions undefined for some values in their domain?
17. How do the ratios that define the trig functions create the asymptotic behavior of certain trigonometric functions?
18. What is the minimum number of measurements needed to determine all of the measurements in a triangle? Explain your reasoning.
19. Which known characteristics of a triangle are needed to prove the law of sines?
20. What makes a trig function even or odd?

Topic 3: Geometric Reasoning/ Measurement and Units

General

- ❏ How can geometry help us make sense of our world?
- ❏ What are the three basic undefined terms in geometry? Why?
- ❏ How does a definition differ from a postulate? From a theorem?
- ❏ Why is it important to use counterexamples when writing definitions?
- ❏ What is a common mistake people might make when using a protractor?
- ❏ What is a benefit to using a full circle protractor?
- ❏ What is the difference between a sketch, a drawing, and a construction in geometry?
- ❏ Why is the study of circles important to geometry?
- ❏ How do geometric models describe spatial relationships?
- ❏ How could you use visualization, spatial reasoning, and geometric modeling as strategies to help you in your problem-solving?

Subtopic Specific

- ❏ **Transformations and Congruence**

  1. What does it mean for two figures to be congruent?
  2. How is congruence related to translation?
  3. How can you determine whether figures are congruent through your reasoning about rigid transformations?
  4. How is SSS congruence different from SAS congruence?
  5. If a side of one triangle is congruent to a side of another triangle, what information about their angles would allow you to prove the triangles congruent?
  6. What is the difference between congruency and similarity?
  7. How do mathematicians communicate ideas about geometric transformations?
  8. How can you change a figure's position without changing its size and shape?
  9. How do you describe isometry in terms of transformations?
  10. What is an example of a rigid motion or a sequence of rigid motions that will transform one congruent figure onto another congruent figure?
  11. How can congruent triangles be used to solve problems?

12. How can you change a figure's position without changing its size and shape?
13. What is the relationship between a translation and a rigid motion?
14. How do you describe the properties of a translation, and its effect on the congruence and orientation of figures?
15. Based on an image and pre-image, how can we identify when and what type of transformation has taken place?
16. How does the dilation factor affect the perimeter and area of a figure? Be specific.
17. What is the difference between dilations and vertical stretch/compressions?

❏ **Similarity**

1. What does it mean for two objects to be similar?
2. What is the relationship between congruence and similarity?
3. How can similarity be modeled in real-life situations?
4. What is meant by self-similarity?
5. When two triangles are similar, how can you use their similarity to solve applications?
6. How are the side lengths, perimeters, and areas of similar rectangles related?
7. What is the relationship between the geometric mean and similarity?
8. How can similarity between figures be proven?
9. How are the ratio of the perimeters and the ratio of corresponding sides for similar polygons related?
10. How does comparing similar polygons describe the relationship between them?

❏ **Right Triangles and Trigonometry**

1. The word trigonometry is based upon what common geometric figure? Why?
2. What is trigonometry?
3. What is the connection between the unit circle coordinates and the associated sine and cosine values? Explain.
4. What is the relationship between a tangent line to a circle and the tangent function?
5. How does tangent relate to sine and cosine?
6. What is the relationship of sine, cosine, and tangent to both the unit circle on a coordinate plane and a right triangle?

7. What are the domain and range for the three basic trigonometric functions?
8. How do the graphs of the trigonometric functions support the idea of periodicity?
9. What is the relationship between radians and degrees?
10. What are at least two different real-world applications for which trig is a useful tool?
11. How does the angle of elevation differ from that of the angle of depression?
12. Why might knowing how to determine an inverse trig function be helpful in solving a word problem?
13. What happens if you attempt to evaluate an inverse trigonometric function at an interval that is not in its domain? Why?
14. Why are there asymptotes in the graph of the tangent function? Where are the asymptotes located?
15. How do you determine the period and amplitude of a sine or cosine function and what do they tell you about the function?
16. How would you describe the behavior of the basic sine curve, the basic cosine curve, and the basic tangent curve?
17. How do you represent a reciprocal trigonometric function on your graphing calculator?
18. What are the two different notations of an inverse trig function?
19. If you start with $y=\sin x$, how does a phase shift of π affect the graph? How about a phase shift of $-\pi$?
20. How do the graphs of $y=\sin(x)$, $y=\sin(2x)$, and $y=\sin(0.5x)$ differ?

❏ **Circles**
1. How can circles be used to describe our world? Cite one specific example.
2. What would the circumference be called if you were looking at a square or triangle instead of a circle?
3. How is the area of a circle similar to the area of a parallelogram?
4. How is a central angle related to the arc it intercepts?
5. What does it mean for an angle to be subtended by an arc?
6. How are angles and intercepted arcs of circles related and applied?
7. If you inscribe a circle in a square, how can you figure out the area of the portion of the square that is outside the circle?
8. How does the number of degrees in the central angle relate to the area of the sector formed by the central angle and the circle?
9. How many tangents can a single circle have?

10. When will two lines tangent to the same circle not intersect? Two secants?
11. Will any diameter of a circle be perpendicular to a tangent at a point of intersection along the circumference of a circle? Explain.
12. Why do you think GPS satellites circle in six orbital planes?
13. Why do you think that right angles play such an important role in the arc-angle relationships in circles?
14. How is the distance formula related to the equation of a circle?
15. How is the Pythagorean theorem related to the equation of a circle?
16. Would the equation of a circle be classified as a function. Why or why not?
17. How would you go about proving that a given point was actually the center of a circle?
18. How can equations of circles be applied to sports?
19. What is the difference between a circumscribed circle and an inscribed circle?
20. How is the orthocenter different from the circumcenter?
21. How is the centroid different from the other centers of the triangle?
22. If a general equation form was given for the circle, what strategy would have to be used to determine the center of a circle and the radius?

❏ **Proof, Logic, and Reasoning**

1. What are some similarities and differences between inductive and deductive reasoning?
2. How are counterexamples used in deductive reasoning?
3. How can you determine whether a conclusion is valid?
4. How can you explain the proof in your own words?
5. Why do you think proof, in mathematics, is important? Support your answer.
6. What are some of the different types of proof and when would they best be used?
7. How does the use of geometric constructions parallel the sequential development of proofs?
8. How is coordinate proof related to both geometric and algebraic reasoning? Be specific.

❏ **2-D and 3-D Geometric Measurement**

1. How do measuring and labeling units help us make sense of our world? Be specific.
2. Why can different units represent the same measurement?

3. When is a solid a polyhedron?
4. What is the difference between a vertex of a pyramid and the apex of a pyramid?
5. How can you determine the intersection of a solid and a plane?
6. Are the diameter and the radius of a sphere just like the diameter and the radius of a circle? Why or why not?
7. The surface area of a particular sphere (in square units) is equal to the volume of the sphere (in cubic units). What, if anything, can you learn about the sphere?
8. How do the dimensions of a sphere affect its volume?
9. What is the relationship between the area of a circle and the volume of a sphere?
10. How is algebraic reasoning used in applications involving geometric formulas and contextual problems?
11. How does knowledge of two-dimensional figures assist in solving problems involving 3-D figures?
12. What type of function is surface area? Why?
13. What type of function is volume? Why?
14. What is the ratio that relates the surface areas of similar solids?
15. What is the ratio that relates the volumes of similar solids?

❏ **Vectors**

1. How are a vector and a ray similar? Different?
2. How can you determine if two vectors are parallel?
3. How is the resultant vector affected if the direction of one of the vectors is reversed?
4. How is working with vectors and scalar multiplication geometrically on a plane similar to working with slope?
5. If two vectors are parallel and one of the vectors is multiplied by a scalar, do the vectors remain parallel?
6. How does the unit vector differ from a general vector?
7. What are the benefits of abstractly representing objects and forces in action?
8. How can you determine situations for which an algebraic understanding of vectors is more beneficial than a geometric understanding (and vice versa)?
9. How would you describe the unit vector, its notation, and the role it plays when working with vector applications?
10. How can the components of a vector can be compared to shadows?
11. Vectors can be broken down into what two components? Describe them.

Topical Questions

12. What role does trigonometry play in determining the components of a vector?
13. How do you use vectors to show resultant direction due to wind or water currents?
14. What is meant by the term reference point and why is it important?
15. What is meant by the term resultant vector?
16. How does the location and direction of a vector impact the writing of its components?

Topic 4: Data Analysis, Probability, & Statistics

General

- ❏ What are some ways that data is being collected on you every day?
- ❏ What are some ways that data is used?
- ❏ How do you think the use of modern computers is transforming statistics and the process of scientific discovery?
- ❏ How do you think statistics has changed in the last 20 years?
- ❏ How often do you encounter statistics in your daily life?
- ❏ What are the advantages and disadvantages of analyzing data by hand versus using technology?

Subtopic Specific

- ❏ **Data Analysis and Statistics**
    1. How does the term "variable" in statistics compare with how the term variable is used in algebraic reasoning?
    2. What does it mean for data to be distributed?
    3. What are examples of situations where you would want to compare two different distributions?
    4. How would you compare the variability of two different distributions?
    5. How can each of the three centers for a set of data be visualized on a dot plot?
    6. How does the context of the problem inform the type of center you choose to use?
    7. How do the different graphs emphasize different aspects of the data? What is an outlier? How might you identify one?
    8. Will data with a larger range always have a higher standard deviation? Explain.
    9. What would a standard deviation of zero mean?
    10. Why is a normal distribution sometimes called a bell curve and how does the slope of the curve differ along different points?
    11. How did you approximate the area under the curve and what do you think the sum of the area under the entire curve should be? Why?
    12. What is bivariate data and how do two-way tables involve bivariate data?
    13. How could you visualize a two-way frequency table graphically?
    14. How can relative frequency tables be used to summarize data?

15. What are some differences between a joint relative frequency and a marginal relative frequency?
16. Does the order in which you plot points on a scatterplot matter? What does each point/dot represent on the scatterplot? Explain.
17. Do you think perfect positive correlation and negative correlation exist often between variables? Explain.
18. Do lines of best fit have to pass through all the points? Explain why or why not.
19. Is it possible to have more than one line of best fit? Explain.
20. How did you determine the window for looking at your graphs on a graphing calculator? How do you think the scale chosen might affect how you perceive the fit?
21. What are the benefits of performing a linear regression analysis?
22. How can the slope of a linear regression equation be interpreted?
23. How can the correlation coefficient of a data set be interpreted?
24. What is the accuracy of a prediction obtained from a linear regression equation?
25. When using a linear regression equation to make predictions, what mistakes can be made?
26. How do we make predictions and informed decisions based on current numerical information?
27. Does correlation imply causation? Explain.
28. What is the difference between prediction and inference?
29. How can random sampling be used to draw inferences about a population?
30. How does the degree of visual overlap in two numerical data distributions with similar variabilities lead to making a comparative inference about the two populations?

❑ **Probability**

1. How are probability and chance related?
2. What is the difference between theoretical and experimental probability?
3. What is an example of an event in which probability must be determined empirically rather than theoretically?
4. What is the difference between independent and dependent events?
5. What is the difference between conditional probability and simple probability?
6. What is the difference between: P(A | B) and P(A and B)?
7. How can you tell if an event is mutually exclusive or not?
8. How can a sample space be represented with a list?

9. How is representing a sample space different than finding the possible outcomes?
10. When is an instance when using the fundamental counting principle is more advantageous than using a tree diagram? Why?
11. What question can help you decide if you need to use a permutation or a combination?
12. How does an outcome table relate to the tree diagram it represents? Be specific.
13. Can probability be used to predict future events? Explain.
14. What is a simulation? How can it be useful?
15. What is the difference between using the fundamental counting principle and using the formulas for permutations and combinations?
16. How can permutations and combinations be useful in sports?

CHAPTER 6

Writing About...

Writing About is a small group writing activity that can be used strategically to support students who struggle with writing, particularly language learners. Just because a student can verbally tell you something does not mean that they can write that same response and support it with evidence. Prior to this activity, you might invite the ELA teacher to visit the class and share what makes a good paragraph so common expectations can be set that support the work in ELA.

Begin by giving students two or three index cards or scraps of paper. Students are to study the word cloud and write one or two sentences about the topic using words they find in the word cloud. Each student shares their sentences with the group and together creates a paragraph about the topic. The index cards allow students to sequence the sentences to build a thoughtful and complete paragraph. They combine similar sentences, check for an introduction, conclusion, etc. This provides an opportunity for students to practice building a paragraph about a topic. As students first work in a group of three or four, they can then begin to work with a smaller group or a partner. The activity can be extended later as an individual writing activity as students are developing stamina for writing. Be aware that not all students will progress at the same pace.

Extensions: Students can sort the words found in the word clouds and create a mapping. Students can work in small groups, pairs, or individually. Students need to be able to articulate their sorting/mapping rule. If students are doing a mapping, they can draw connectors, use string/yarn, or use something like WikkiStix™. If using WikkiStix™, be sure students are working on a piece of construction paper or scrap paper that will not matter if the sticky gets on it or not.

Note: The word cards are grouped by topic, as different schools offer different courses. And, the words are not alphabetical, but are grouped more by level of typical course.

Once students sort their word set and show their connections, they need to write down their sorting rule in their Mathematician's Notebook. Once all groups are finished, students can do a Walk About Review where they observe the other groups' mapping/sorting and make notes about what they think their sorting rules were. Then, the whole group can come back together and discuss what they observed. Some questions that you might use to facilitate the discussion could include the following:

❏ What were the similarities you observed between the mappings?
❏ What were some differences?
❏ Were you able to identify the correct sorting rule for the other groups? Why or why not?

Suggested directions for the mapping/sort:

Study the words. Sort the words. Sort the sets of words that seem to go together. You may use your string/WikkiStix™ to show connections between the words. Explain your sorting rule fully. If directed, create a second sorting with a different rule.

Suggested directions for the Walk About Review:

As you walk about and review the other groups mappings, do not talk, look over the mapping and in your notebook identify what you think the groups' sorting/mapping rule is and why. You will have a set amount of time at each mapping, so use it wisely and efficiently.

Ideas for Display: Groups can create a graffiti board using chart paper to capture their paragraph. These group boards can then be put together to create a graffiti wall. The class could do a gallery walk to view what was developed, provide feedback, and/or reflect on the process.

Writing About...

Writing about...

Study the word cloud below. Create at least two statements about absolute value using the key words you see in the word cloud. With your group, use your sentences to create a paragraph about absolute value.

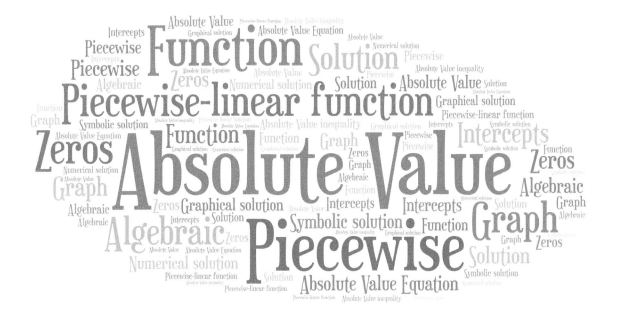

Authentic Opportunities for Writing about Math in High School

Writing about...

Study the word cloud below. Create at least two statements about Algebra 2 using the key words you see in the word cloud. With your group, use your sentences to create a paragraph about advanced algebra.

Writing about...

Study the word cloud below. Create at least two statements about the Cartesian Coordinate plane and plotting points using the key words you see in the word cloud. With your group, use your sentences to create a paragraph about plotting points.

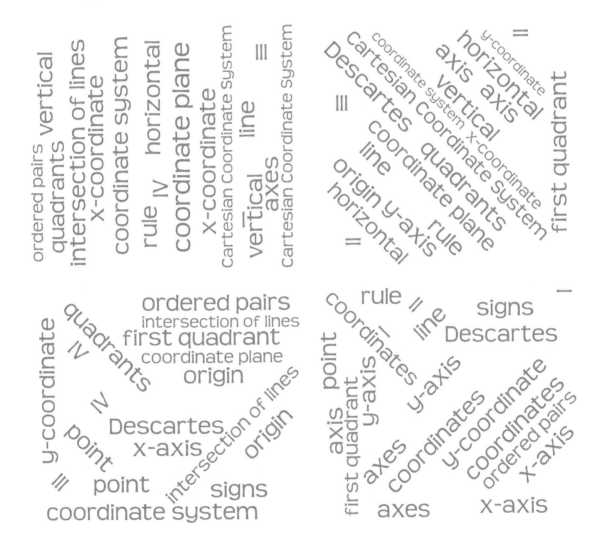

Authentic Opportunities for Writing about Math in High School

Writing about...

Study the word cloud below. Create at least two statements about circles using the key words you see in the word cloud. With your group, use your sentences to create a paragraph about circles.

Writing about...

Study the word cloud below. Create at least two statements about conic sections using the key words you see in the word cloud. With your group, use your sentences to create a paragraph about conic sections.

Writing about...

Study the word cloud below. Create at least two statements about data analysis using the key words you see in the word cloud. With your group, use your sentences to create a paragraph about data analysis.

Writing About...

Writing about...

Study the word cloud below. Create at least two statements about exponential and logarithmic functions using the key words you see in the word cloud. With your group, use your sentences to create a paragraph about exponential and logarithmic functions.

Horizontal Line Test
Natural Exponential Function
Properties of Logarithms
Logistic Model **Exponent**
Arithmetic Operations Common Logarithm
Inverse Functions
One-to-One Function
Models
Arithmetic Operations Exponential Function
Exponential Function Horizontal Line Test
Exponent **Exponent** Properties of Logarithms
Exponential Function Natural Logarithm General Logarithm **Logarithm**
Models Combining Functions Continuously Compounded Interest
Logarithm Common Logarithm **Composition of Functions**
Models Inverse Functions **Logarithmic Function** Continuously Compounded Interest
One-to-One Function Natural Logarithm Natural Exponential Function
General Logarithm Common Logarithm Composition of Functions Exponent
Common Logarithm Compound Interest Combining Functions
Combining Functions Solving One-to-One Function
Natural Logarithm **Inverse Functions**
Common Logarithm **One-to-One Function**
Natural Exponential Function **Compound Interest**
Compound Interest Inverse Properties
Natural Exponential Function Horizontal Line Test
Compound Interest Arithmetic Operations Logarithm Composition of Functions **Solving**
Arithmetic Operations Combining Functions Inverse Properties
General Logarithm **Solving** Logistic Model
Exponential Function Inverse Properties Logarithmic Function
Logarithmic Function Continuously Compounded Interest **Solving**
Combining Functions **Exponent** Compound Interest
Horizontal Line Test Logarithmic Function Logistic Model
Arithmetic Operations Properties of Logarithms Logarithm
General Logarithm Inverse Properties Models
Logarithmic Function **Logarithm**
Exponential Function General Logarithm
Properties of Logarithms **Logistic Model**
Continuously Compounded Interest
Inverse Functions
Composition of Functions
Composition of Functions
Natural Logarithm

65

Authentic Opportunities for Writing about Math in High School

Writing about...

Study the word cloud below. Create at least two statements about exponential functions using the key words you see in the word cloud. With your group, use your sentences to create a paragraph about exponential functions.

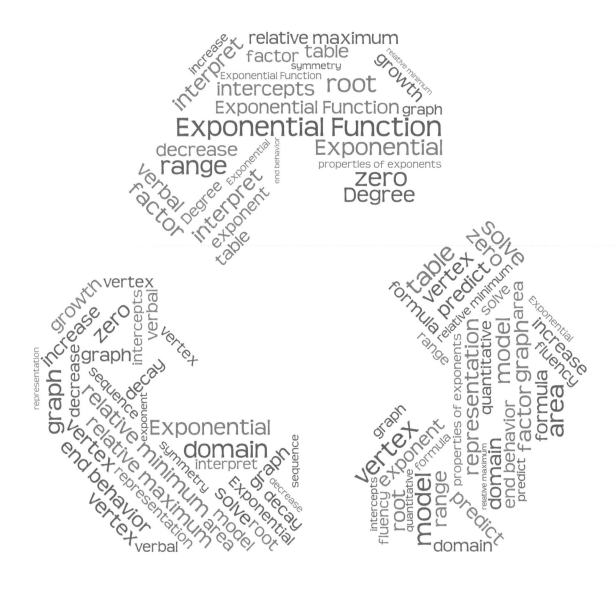

66

Writing about...

Study the word cloud below. Create at least two statements about expressions and equations using the key words you see in the word cloud. With your group, use your sentences to create a paragraph about expressions and equations.

Authentic Opportunities for Writing about Math in High School

Writing about...

Study the word cloud below. Create at least two statements about geometric measurement using the key words you see in the word cloud. With your group, use your sentences to create a paragraph about geometric measurement.

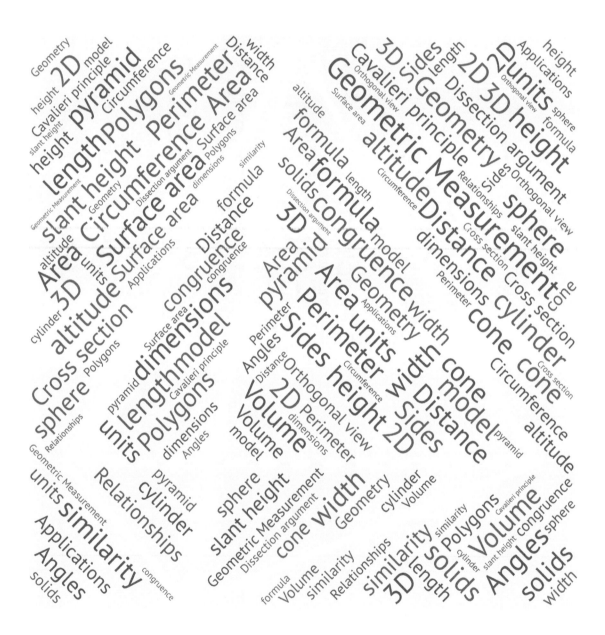

Writing about...

Study the word cloud below. Create at least two statements about 3D geometry using the key words you see in the word cloud. With your group, use your sentences to create a paragraph about the 3D geometry.

Authentic Opportunities for Writing about Math in High School

Writing about...

Study the word cloud below. Create at least two statements about geometry using the key words you see in the word cloud. With your group, use your sentences to create a paragraph about geometry.

70

Writing About...

Writing about...

Study the word cloud below. Create at least two statements about similarity using the key words you see in the word cloud. With your group, use your sentences to create a paragraph about similarity.

Authentic Opportunities for Writing about Math in High School

Writing about...

Study the word cloud below. Create at least two statements about triangles using the key words you see in the word cloud. With your group, use your sentences to create a paragraph about triangles.

Writing About...

Writing about...

Study the word cloud below. Create at least two statements about inequality using the key words you see in the word cloud. With your group, use your sentences to create a paragraph about inequality.

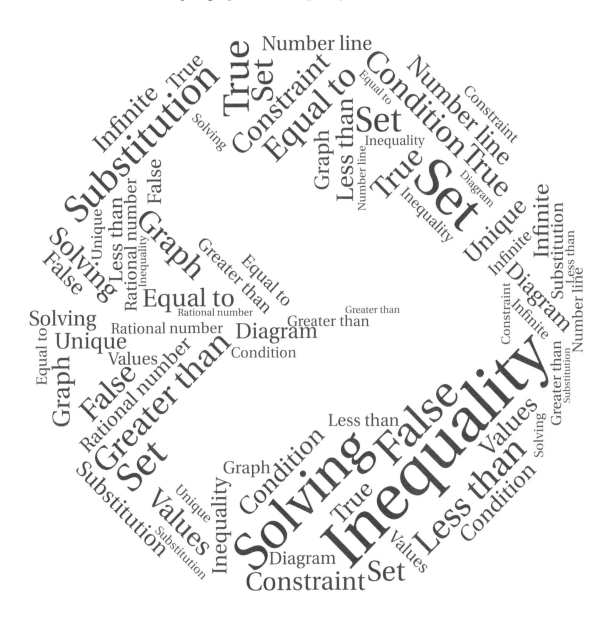

Authentic Opportunities for Writing about Math in High School

Writing about...

Study the word cloud below. Create at least two statements about limit properties using the key words you see in the word cloud. With your group, use your sentences to create a paragraph about limit properties.

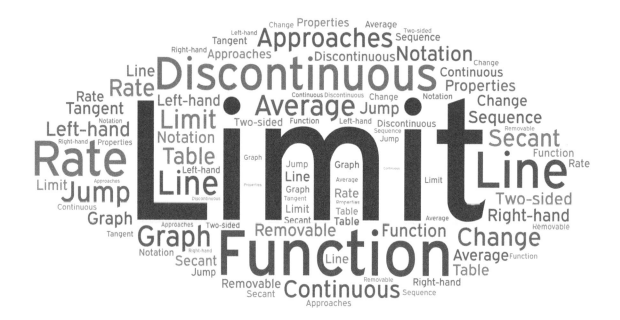

Writing about...

Study the word cloud below. Create at least two statements about linear functions using the key words you see in the word cloud. With your group, use your sentences to create a paragraph about linear functions.

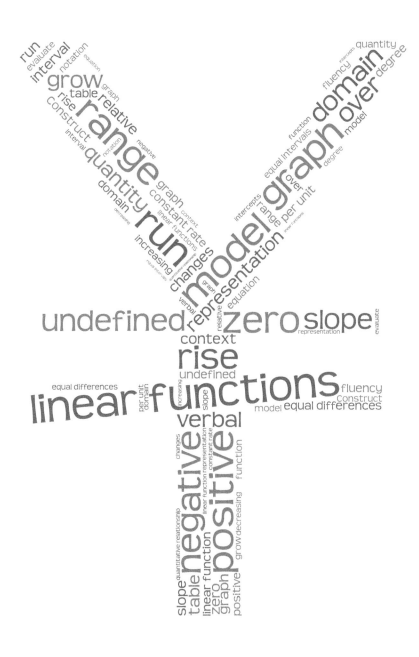

Authentic Opportunities for Writing about Math in High School

Writing about...

Study the word cloud below. Create at least two statements about modeling using the key words you see in the word cloud. With your group, use your sentences to create a paragraph about modeling.

Writing About...

Writing about...

Study the word cloud below. Create at least two statements about polynomial and rational functions using the key words you see in the word cloud. With your group, use your sentences to create a paragraph about polynomial and rational functions.

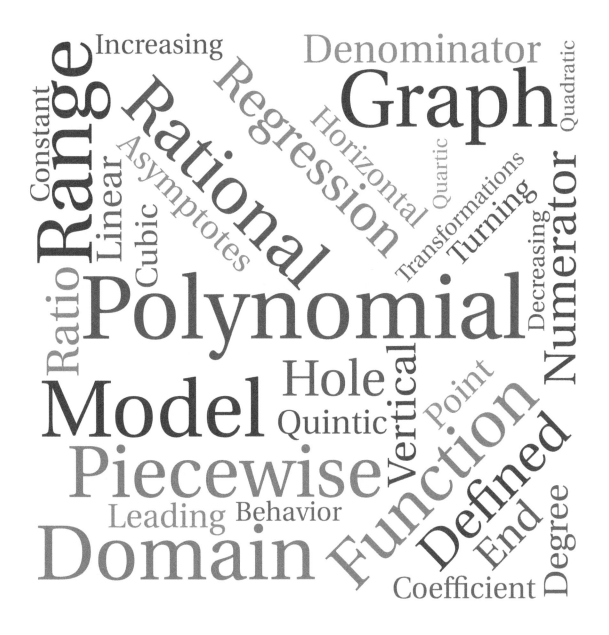

Authentic Opportunities for Writing about Math in High School

Writing about...

Study the word cloud below. Create at least two statements about probability using the key words you see in the word cloud. With your group, use your sentences to create a paragraph about probability.

Writing about...

Study the word cloud below. Create at least two statements about the Pythagorean Theorem using the key words you see in the word cloud. With your group, use your sentences to create a paragraph about the Pythagorean Theorem.

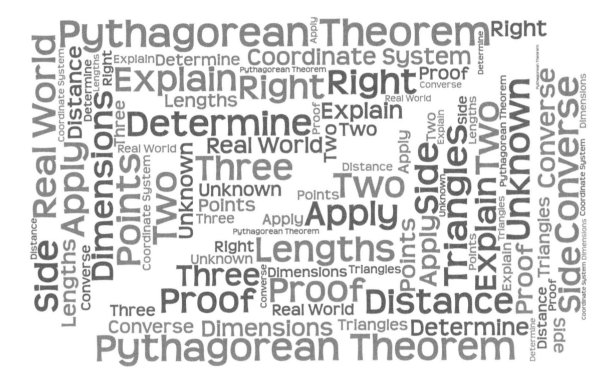

Authentic Opportunities for Writing about Math in High School

Writing about...

Study the word cloud below. Create at least two statements about quadratic functions using the key words you see in the word cloud. With your group, use your sentences to create a paragraph about quadratic functions.

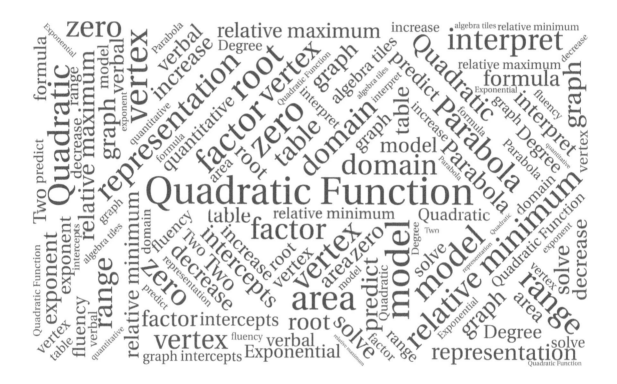

Writing About...

Writing about...

Study the word cloud below. Create at least two statements about the Real Number System using the key words you see in the word cloud. With your group, use your sentences to create a paragraph about the Real Number System.

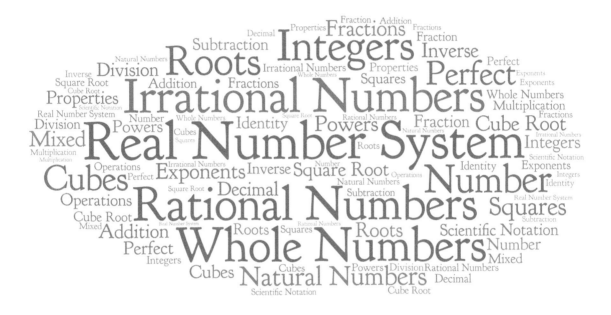

Authentic Opportunities for Writing about Math in High School

Writing about...

Study the word cloud below. Create at least two statements about right triangle trigonometry using the key words you see in the word cloud. With your group, use your sentences to create a paragraph about right triangle trigonometry.

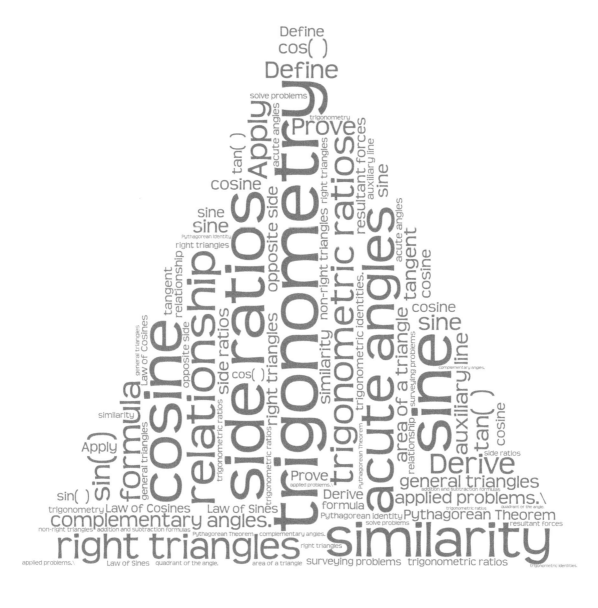

Writing about...

Study the word cloud below. Create at least two statements about statistics using the key words you see in the word cloud. With your group, use your sentences to create a paragraph about statistics.

Authentic Opportunities for Writing about Math in High School

Writing about...

Study the word cloud below. Create at least two statements about systems and matrices using the key words you see in the word cloud. With your group, use your sentences to create a paragraph about systems and matrices.

Writing About...

Writing about...

Study the word cloud below. Create at least two statements about the tools of mathematics using the key words you see in the word cloud. With your group, use your sentences to create a paragraph about the tools of mathematics.

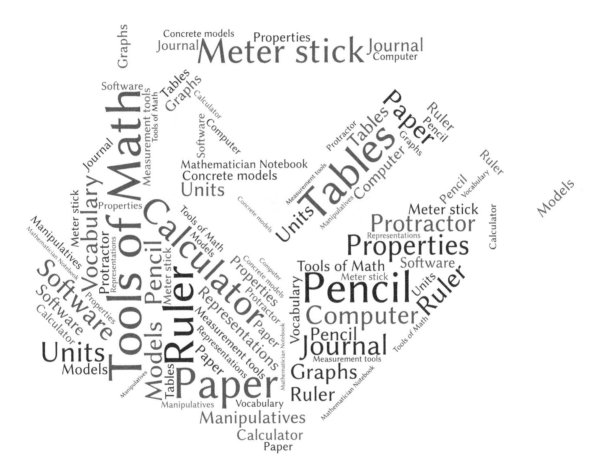

Authentic Opportunities for Writing about Math in High School

Writing about...

Study the word cloud below. Create at least two statements about transformational geometry using the key words you see in the word cloud. With your group, use your sentences to create a paragraph about transformational geometry.

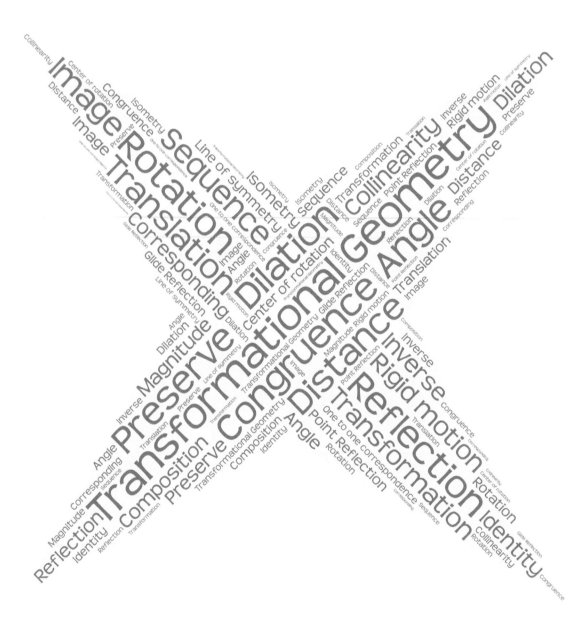

Topic 1: Number and Quantity

Absolute value	**Coefficient**
Constant term	**Constraint**
Degree of a term	**Irrational number**

Rational number	Real number system
Zero product property	Arithmetic sequence
Complex numbers	Conjugate pair

Writing About...

Geometric sequence	**Imaginary numbers**
Logarithm	**Natural logarithm**
Polynomial expression	**Vector**

Topic 2: Algebraic Reasoning

Average Rate of Change	**Completing the square**
Decreasing (function)	**Dependent variable**
Domain of a function	**Equivalence**

Writing About...

Exponential function	**Factored form of a quadratic**
Function	**Function Family**
Function notation	**Increasing (function)**

Independent variable	**Inverse function**
Linear Function	**Linear terms**
Maximum/ minimum of a function	**Piece-wise function**

Quadratic function	**Quadratic term**
Range of a function	**System of equations**
System of inequalities	**Vertex of a graph**

x-intercept	y-intercept
Zero of a function	Amplitude of a periodic function
End behavior of a graph of a function	Even function

Writing About...

Odd function	Asymptote
Horizontal Asymptote	Vertical Asymptote
Identity Function	Logarithmic function

Multiplicity of roots of a polynomial function	**Polynomial function**
Periodic function	**Period of a periodic function**
Pythagorean Identities	**Trigonometric identities**

Writing About...

Rational function	Relative maximum of a function
Relative minimum of a function	Unit circle

97

Topic 3: Geometric Reasoning/ Measurement and Units

Altitude of a figure	**Angle bisector**
Apex	**Arc of a circle**
Auxiliary line	**Axis of rotation**

Central angle of a circle	Chord of a circle
Circumcenter	Circumscribed figure
Complementary angles	Conjecture

Converse of a statement	Negation of a statement
Inverse of a statement	Contrapositive of a statement
Cosine ratio	Sine ratio

Writing About...

Tangent ratio	**Trigonometric ratios**
Cross-section of a 3-D figure	**Geometric transformations**
Dilation	**Reflection**

Rotation	Translation
Isometry	Rigid transformation
Directrix of a parabola	Focus of a parabola

Writing About...

Parabola	Locus of points
Incenter	Orthocenter
Centroid	Euler's Line

Inscribed angle	**Inscribed figure**
Median of a figure	**Perpendicular bisector**
Degree measure of an angle	**Radian measure of an angle**

Writing About...

Vector	**Sector of a circle**
Similar figures	**Tangent line to a circle**
Theorem	**Postulate**

Topic 4: Data Analysis, Probability and Statistics

Association	**Bell-shaped Distribution**
Bimodal distribution	**Categorical data**
Categorical variable	**Causal relationship**

Writing About...

Correlation coefficient	**Normal Distribution**
Five-number data	**Model**
Numerical data	**Outlier**

Relative Frequency Table	Residual
Standard deviation	Statistic
Statistical question	Two-way table

Margin of error	**Observational study**
Random	**Sample**
Population	**Statistical survey**

Addition rule for probability	Conditional probability
Dependent events	Independent events
Geometric probability	Probability

Writing About...

Permutation	**Combination**
Expected Value	

CHAPTER 7

Journal Prompts

As mentioned previously in the Compare and Contrast section, two of the main components of the Mathematician's Notebook are the glossary and the journal. Journals are a great way for students to keep track of their mathematical journey as well as giving insight into how they think about math and how they have developed and grown over the course of the semester or year. Journals can be a place where students engage with quotes, historical connections to topics, famous people related to the topic of study, misconceptions, and "What if?" scenarios. If you keep a "parking lot" in your classroom for students to list issues they had with the homework from the day before, if only a couple of people had trouble with say problem #2, we might just have other students work that at the board and then see what follow-up questions, if any, were needed. But if ten students had issues with problem #2, then as the teacher, I must ask, "What did I not do that allowed students access to beginning the problem." So, that problem might go into the journal with a discussion and notes around issues students were having.

Journal entries are best assessed separately from the rest of the Mathematician's Notebook. Even a basic *All, Most, Some, None* format works well. Journals are a place where you can dialogue with students about topics and engage with them in a different way. The authors never wrote directly on the student's notebook pages, as that was their own work, but wrote on sticky notes and attached it to the page(s) where comments were appropriate.

Journal Prompts

Following are some beginning suggestions for journal prompts that can be used throughout the year. Some are very focused around mathematical topics, and some are just for fun for students to allow their imagination to run wild. Hopefully, these will serve as the basis for you to add many additional ideas of your own.

Math Specific

Math-ography: Students write about their prior experiences with mathematics. No comments on specific teachers are allowed. What topics made sense, what topics were challenging, and how do they see mathematics fitting into their occupational plans.

WRITE YOUR "MATH-OGRAPHY"

Include your:

- earliest remembrances of counting, learning about numbers,
- elementary school work, topics you "got" as well as topics that were challenging,
- your goals for this class, and
- what you see yourself doing after you complete high school and what role math will play in that….

DO NOT name teacher names!

A self-evaluation: Usually assigned around the first interim period for grading, this provides the opportunity for students to reflect on their work so far. It has proved helpful to give students a list of questions to guide their reflections. Some examples are as follows:

- ❏ Do you have a dedicated place at home to study?
- ❏ Have you been regular in your attendance?
- ❏ How would you rate your engagement in class when you are here?
- ❏ How much time outside of class are you spending on work for this class?
- ❏ Do you feel that you are getting enough sleep and rest?
- ❏ How are your eating habits affecting your schoolwork?

Authentic Opportunities for Writing about Math in High School

Writing to Explain

Option 1: Students write an explanation for a student in their class who was absent the day they learned about/how to (insert topic/activity/procedure for the day here.)

Option 2: Students area assigned a mathematical topic such as one of the geometric transformations or measures of central tendency. Then they complete the following.

- ❏ You are an exponential function. Tell us everything we should know about you and your family (substitute any function family).
- ❏ You are a normal distribution of data. Tell us everything we should know about you.
- ❏ You are a complex number. Tell us everything we should know about you.
- ❏ You are an isometry. Tell us everything we should know about you.
- ❏ You are one of a conjugate pair. Tell us everything we should know about you.

Creative Writing

Students complete each of the following prompts using their imagination. Encourage students to just not write but to also use drawings and sketches and color as they complete the prompt.

- ❏ If I were a function, I would be _____ because....
- ❏ If I were a geometric postulate or theorem, I would be _____ because....
- ❏ You wake up tomorrow morning and find that circles no longer exist. How do you get to school? Be sure to sketch the road, vehicle, etc., that you would use. Extension: How else will circles not existing impact your life? Be specific.
- ❏ If I were a mathematical pattern, I would be _____ because....
- ❏ What I find the most challenging with _____ (current topic) is.... Explain why.
- ❏ When I see a math problem with words, I feel _____ because....
- ❏ Choose a character from literature and describe how he/she might use mathematics in what he/she does.
- ❏ Write a compliment to yourself for something you accomplished in math recently.

Journal Prompts

Describe: Students free write for either a specified amount of time (Timed writing) on a particular assignment or topic, or for a particular amount of writing – a paragraph, 5–6 sentences, etc.

- ...what you see in a picture prompt
- ...the steps in a process – such as creating a specific geometric construction
- ...the difference in the types of slope of linear functions.
- ...why can't you divide by zero.
- ...what the difference is between 0.999... and $\frac{1}{2}+\frac{1}{4}+\frac{1}{8}+...$
- ...what caused you confusion in the homework.
- ...how the slope-intercept form of a line supports the understanding of transformations in quadratic functions.
- ...what it means to "find a solution" to a quadratic equation.
- ...the importance of graphical representations of data.
- ...how you know if a system of equations has an infinite number of solutions.
- ... a contextual scenario where the linear identity function would be an appropriate model.

Research: A famous historical mathematician, scientist, etc.

Example: Research M.C. Escher and write at least a one-page overview of his life in your journal. Include **who** he was, **when** he lived, **where** he was from, **what** he is best known for and **how** that relates to this class, and a couple of interesting facts about his life.

Using Quotes

Option 1: Students write the quote, they write what it meant in the time it was written, and then how it would be applicable to them in math class today.

Example: *I apologize for the length of this letter. Had I but more time it would have been shorter.* – Blaise Pascal

- Students copy the quote in their journal.
- They then write a couple of sentences about what they think Pascal meant in light of his time period. They may need to ask their history or English teacher for help here. Back in Pascal's time, people wrote drafts of their letters, much as students write drafts for their English papers today. So, he did not have a lot of time to "clean up" his letter and make it shorter.

❏ Students conclude by writing a couple of sentences about how this will apply to their work in math class. When writing in mathematics, in their Mathematician's Notebook, etc. students need to learn to not only be precise, but to be succinct with their explanations and to the point.

Option 2: Students write a response to the author.

Option 3: Students write about what questions the quote prompts them to think about.

Option 4: The students describe what the quote means to them.

Below is a beginning list of quotes which span from historical to modern day, includes a diverse group of individuals, and cuts across disciplines to include the humanities as well as the sciences.

The purpose of models is not to fit the data but to sharpen the questions. – Samuel Karlin[1]

Any fool can know. The point is to understand. – Albert Einstein[2]

What is mathematics? It is only a systematic effort of solving puzzles posed by nature. – Shakuntala Devi[3]

Mathematics knows no races or geographic boundaries; for mathematics, the cultural world is one country. – David Hilbert[3]

Math proficiency is the gateway to a number of incredible careers that students may never have considered. – Danica McKellar[1]

Simple laws can very well describe complex structures. The miracle is not the complexity of our world, but the simplicity of the equations describing that complexity. – Sander Bais[4]

Life without geometry is pointless. – Unknown[5]

We learn more by looking for the answer to a question and not finding it than we do from learning the answer itself. – Lloyd Alexander[5]

You don't understand anything until you learn it more than one way. – Marvin Minsky[5]

The most useful piece of learning for the uses of life is to unlearn what is untrue. – Antisthenes[5]

Knowledge is like money: to be of value it must circulate, and in circulating it can increase in quantity and, hopefully, in value. – Louis L'Amour[5]

Euclid's first common notion is this: Things which are equal to the same things are equal to each other. That's a rule of mathematical reasoning and its true because it works. – Abraham Lincoln[8]

It's amazing what one can do when one doesn't know what one can't do. – Garfield the Cat[2]

Number is the within of all things. – Pythagoras[1]

What we know is not much. What we do not know is immense. – Pierre-Simon Laplace[2]

Nature's Great Book is written in mathematical symbols. – Galileo Galilei[2]

If you don't like the answer, ask a different question. – Dr. Larry Fleinhardt – NUMB3RS[6]

Measure what is measurable and make measurable what is not so." – Galileo Galilei[2]

A thing is obvious mathematically after you see it. – R. D. Carmichael[7]

One accurate measurement is worth a thousand expert opinions. – Rear Admiral Grace Hopper[1]

Don't give up on it. Just stick with it. Don't listen to people who always tell you it's hard, and walk away from it. – Annie Easley[8]

Idealism increases in direct proportion to one's distance from the problem. – John Galsworthy, Nobel Laureate in Literature[5]

References

1. https://quotefancy.com/
2. https://www.goodreads.com
3. https://www.prodigygame.com/main-en/blog/math-quotes/
4. https://www.famousscientists.org/magnificent-mathematics-quotes/
5. https://www.livingmath.net/quotes
6. https://www.quotes.net
7. https://www.prodigygame.com/main-en/blog/math-quotes/
8. https://laidlawscholars.network/posts/just-stick-with-it

CHAPTER 8

Poetry/Prose

In the spirit of writing in response to a quote (described in the Journal Prompts), this section begins with one from JoAnne Growney's blog Intersections – Poetry with Mathematics

> Mathematical language can heighten the imagery of a poem; mathematical structure can deepen its effect.

The precision of language required of both disciplines makes the intersection of mathematics and poetry seem almost obvious. Providing the opportunity to see the connection allows students to explore this relationship while deepening their understanding of mathematics and writing skills.

Acrostic: explain – one word, expression, describing the main word…
Example:

Circles

Centers
Inside
Radii and
Chords
Lopsided never
Elegant with perfect
Symmetry

Math

Mysterious to some.
Ability required.
The language of our world.
Hurts my brain!

Beginning List of Terms: (See Word Lists in Compare and Contrast for Additional Terms)

Equation
Expression
Geometry (or any specific shape or term)
Inequality
Math
Numbers (or use any specific number set)
Probability
Proportional
Ratio
Statistics

Fibonacci poem: Students can create their own "Fibonacci" poem where each line of the poem has the number of words as found in the sequence. This can be differentiated for students by having them use only the first three or four numbers found in the sequence or more if their writing skills allow. The topic of the poem can be of their choosing or can be assigned. The following poem models 1, 1, 2, 3, 5, 8.

Triangles

Triangles
Pointed
Three sides
And three angles
Walking from vertex to vertex
Around the corners, either obtuse, acute, or right.

Haiku: has three lines, five syllables in first and third lines and seven syllables in the second line

Example:

Ratios

Ratios compare
Proportional reasoning
Ratios equal

Pi poem: Is similar to the Fibonacci poem. Students write lines based on the digits in pi: 3.1415926535 8979323846 2643383279. This can be differentiated for students by having them use only the first three or four digits found in pi or more if their writing skills allow.

Students can also have fun with the topic as in the example below.
Example:

Pi

My favorite pie --
cherry.
Flakey crust, tart flavor,
Yum!

Free verse/free write: There is no specific form, meter, or rhyme scheme. Students can have the freedom to write as they feel. Some suggestions are given below for types of free writes students may enjoy doing.

Cartoons
Commercial, Infographic, Public Service Announcement (PSA)
Free Verse Poem
Graphic novel, for example, The Adventures of Slope Boy
Historical Fiction
Math Carols and/or seasonal songs
Math words to a current song
Short Story

CHAPTER 9

Cubing and Think Dots

Cubing and think dots are two strategies for differentiation in the classroom. Traditionally students are given a cube with a variety of activities or tasks at a given level. Different cubes can contain different levels of tasks and activities. Think dots work in a similar way. Cards with a certain number of dots are provided as well as a number cube. Students roll the number cube and work the associated activity or task on the card with the corresponding number of dots. You can choose the parameters for your students or create your own set of think dot cards using index cards and practice problems from your chosen curriculum.

As students are working through the various activities, emphasize they are to make statements explaining and supporting their reasoning and thinking, not just "show their work." Students need to be able to use precise mathematical language and symbols in their written work as well as clearly articulate their thinking.

Preparation

Print the cubes and think dot cards on cardstock or heavy paper. Build the cubes carefully, using the dotted lines as the folds. Tape or glue the edges together. Fold the think dot cards and tape together. These resources can be used in a variety of ways in a learning station or as part of intervention support.

Materials List

- Functions Think Dot Cards set
- Function Representation Cube
- Function Type Cube I
- Function Type Cube II
- Function Type Cube III
- Geometric Reasoning Think Dot Cards set
- Geometric Figure Cube
- Transformation Octahedron
- Rotation Degree Spinner
- Number cubes and/or dominoes (Note: You can purchase number cubes and domino sets that show numbers greater than six.)
- Graph paper to use with graphing functions and/or the geometric transformations

Functions

One of the goals of high school mathematics is that students develop functional fluency. What is functional fluency? Functional fluency, at a beginning level, is students looking at any representation of a function and moving seamlessly from one form to another. Students also understand the parent function for the function type and the associated transformations.

(Jones, T. L., April 2015, Functional Fluency blog post, https://tljconsultinggroup.com/functional-fluency/)

In this adaptation of a cubing and think dots activity, there are two cubes. One cube has representations of functions. Another cube has function types. There are a various ways students can engage in the set of six think dot cards.

Option 1: Students can roll the representation cube only and create that representation based upon the function(s) you assign.

Option 2: Students can roll the chosen function cube and the representation cube and create that representation based upon the function they rolled.

Option 3: Students use the cubes in tandem with the think dot cards. The think dot cards are self-explanatory and can be adapted as needed for your students.

Function Representation Cube Options

Numeric/Table: Numeric and tabular forms of a function are the forms usually first encountered by students. They use input and output tables in middle school as well as evaluate equations/functions using substitution. Numerical representations are tables of values that list the input-output pairs for a given function. Students should realize that tables have limitations. They cannot show all the possible inputs for a given function.

Verbal: Verbal representations can be as simple as writing mathematical sentences in words. This can be the beginning point for students. Students should be pushed to be able to interpret the computation(s) performed by the function in context.

Algebraic/Symbolic: Tables have limitations that the algebraic/symbolic representations do not. The function "formula" represents or defines the given function. The symbolic form of a function, while complete and efficient, is limited by not being very visual.

Graph: The graphical representation for a function, on the other hand, is very visual. Students see the input and output pairs plotted on the Cartesian Coordinate System. These representations can first be as scatterplots of the data sets generated by the function. Students then graph functions based upon the generalized function or based upon the parameters set by the context or specific function definition. Graphs, used in conjunction with tables, also help students identify and understand a function's domain and range as well as any restrictions on those. Asymptotes are also easily identified through a table of values on a graphing calculator, even if not shown on the graph itself.

Picture/Diagram: When working with functions such as those geometric functions that are the measurement formulas as well as periodic functions, students often find benefit in creating or using a sketch or diagram. The diagram can show the shape being measured, including the measurements for sides, faces, angles, etc. Diagrams can also support work with trigonometric functions and when working with concepts like vectors and bearings. Students should be able to ultimately create these representations for themselves and not have to rely on always being given a picture or diagram.

Set: Set notation is important when discussing domain and range of a function. Students initially can create a set of ordered pairs from a table of values. Later, once function notation is introduced, students can begin identifying the domain and range for specific functions using interval notation.

Function Type Cubes Options: The following function types and options are included in the three cubes. Determine the cube level needed based upon the function types.

Function Type Cube I

Linear, Absolute Value, Quadratic, and Cubic
　Roll Twice: Students roll the function cube twice and create an example of each writing about the two functions and how they are alike/different.
　Free Choice: Students roll the function cube and choose which function they want to use.

Function Type Cube II

Piecewise-defined, Polynomial, Radical, Rational, Exponential, and Free Choice

Function Type Cube III

Piecewise-defined, Polynomial, Rational, Exponential, Logarithmic, and Periodic

Cubing and Think Dots

**High School Functions
Think Dots Cards**

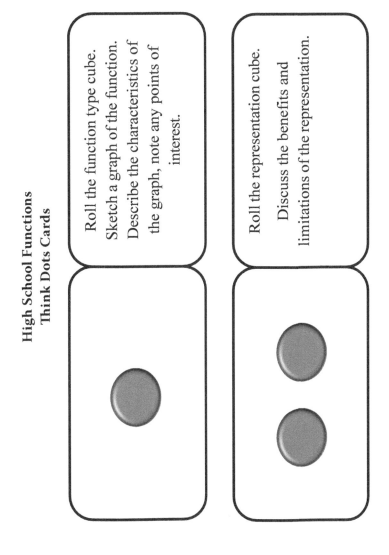

- Roll the function type cube. Sketch a graph of the function. Describe the characteristics of the graph, note any points of interest.

- Roll the representation cube. Discuss the benefits and limitations of the representation.

125

Authentic Opportunities for Writing about Math in High School

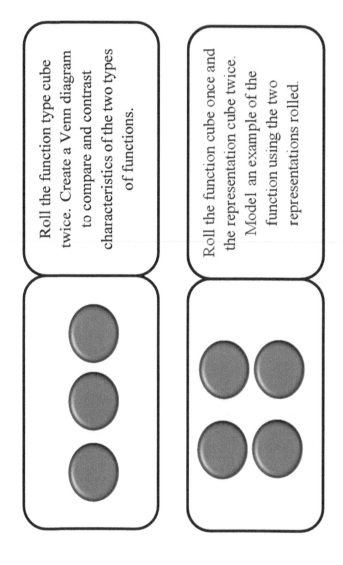

Cubing and Think Dots

Roll the function cube once and the representation cube once. Describe how you would solve a function of the type shown, using the representation shown.

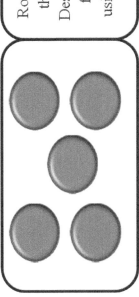

Roll the function cube twice, create an example of each type of function.

Solve the system you created. Discuss the strategy you choose and why.

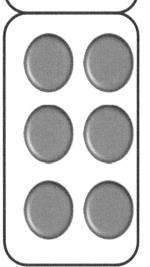

Authentic Opportunities for Writing about Math in High School

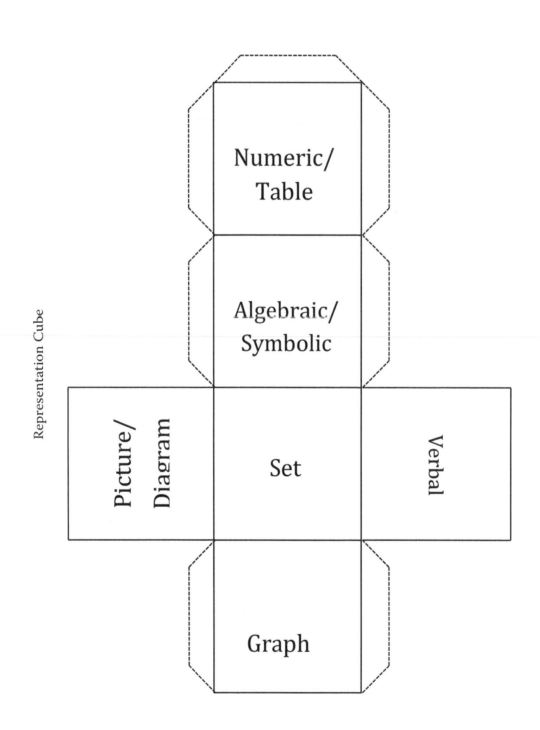

Representation Cube

128

Cubing and Think Dots

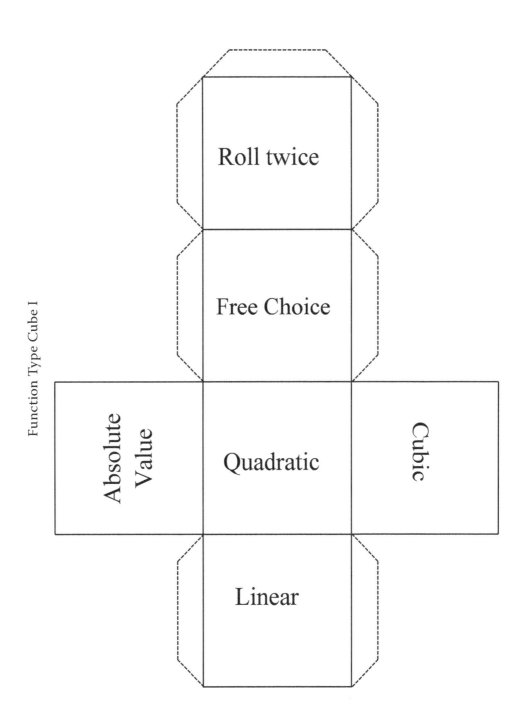

Function Type Cube I

129

Authentic Opportunities for Writing about Math in High School

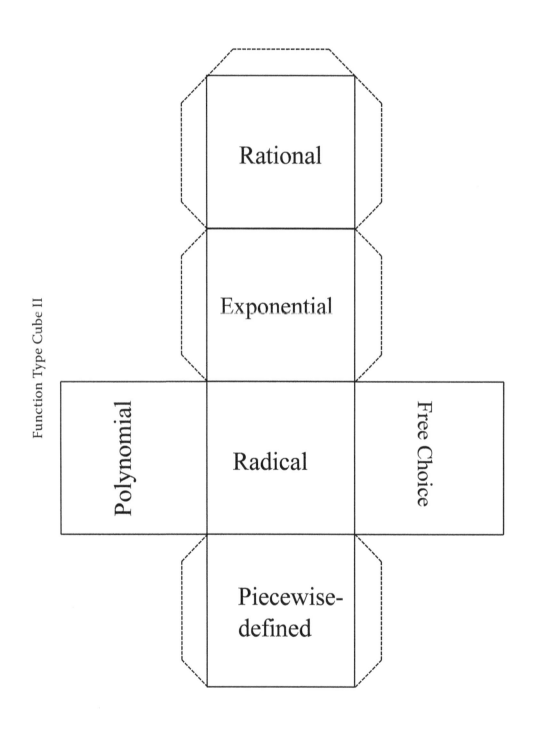

Function Type Cube II

130

Cubing and Think Dots

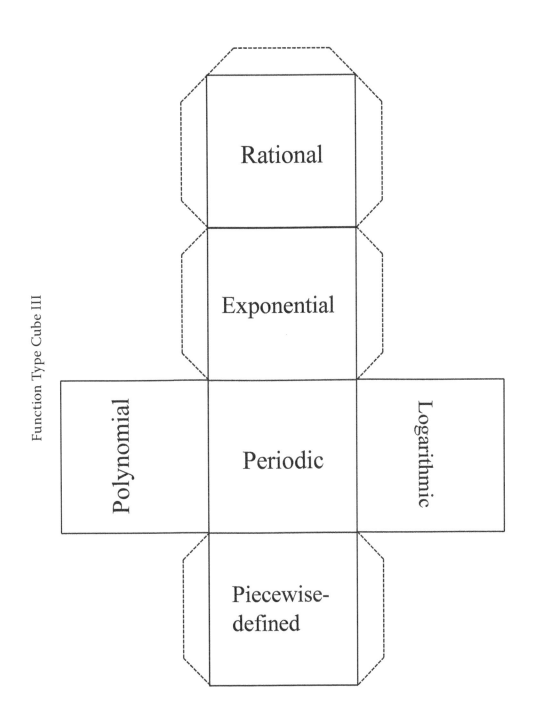

Function Type Cube III

131

Geometric Reasoning

In this adaptation of a cubing and think dots activity, there is a Geometric Reasoning Think Dot Cards set, a Geometric Figure Cube, a Transformation Octahedron, and a Rotation Degree Spinner. Students will also need a standard number cube or a platonic solid die.

Geometric Figure Cube

Students will roll the cube to determine the two-dimensional figure they will be transforming on the Cartesian plane.

Triangle, quadrilateral, hexagon, pentagon, composed figure, and concave figure.

Tetrahedron Transformation Octahedron

Students will roll the octahedron to determine the transformation they will be applying to the figure(s).

Clockwise rotation and Counterclockwise rotation: Students also spin the Rotation Degree Spinner to determine the amount of rotation they perform about the origin.

Translation: Students also roll a number die to determine the amount of translation vertically/horizontally. Using a decahedron, dodecahedron, or icosahedron provides students with quantities greater than 6.

Reflection about the *x*-Axis, *y*-Axis, or *y* = *x*

Expansion dilation and Contraction dilation: Students also roll a number die to determine the positive scale factor for expansion or the negative scale factor for contraction. This is where using a tetrahedron offers students to only use the digits 1 to 4.

Option1: Students can roll just the Geometric Figure Cube and the Transformation Octahedron and perform the transformation based upon their rolls.

Option 2: Students use the cube and octahedron in tandem with the think dot cards. The think dot cards are self-explanatory and can be adapted as needed for your students.

Cubing and Think Dots

Think Dot Cards
Geometry

- Roll the geometric figure cube and sketch the shape on the graph paper.
- Label the vertices with coordinates.
- Roll the transformation cube and transform the pre-image as rolled. Parameters/constraints based on die roll and/or spinner spin.
- Sketch the resulting image and label the vertices with coordinates. Notate the transformation.

•

- Roll the geometric figure cube. Describe the shape in words and sketch the shape on the graph paper.
- Label the vertices with coordinates.
- Roll the transformation cube and transform the pre-image as rolled. Parameters/constraints based on die roll and/or spinner spin.
- Describe the resulting image and label the vertices with coordinates. Notate the transformation.

• •

Authentic Opportunities for Writing about Math in High School

- Roll the geometric figure cube.
- Roll the transformation octahedron two times to create a series of transformations. Sketch each transformation on one piece of graph paper. Parameters/constraints based on die roll and/or spinner spin. Label the vertices.
- Describe in words the resulting transformation(s).

- Roll the geometric figure cube and sketch the shape on graph paper. Label the vertices.
- Roll the transformation octahedron three times.
- Perform the transformations in sequence, sketching each resulting image and labeling the vertices.
- Describe the result of this sequence in words and explain whether it is an isometry.

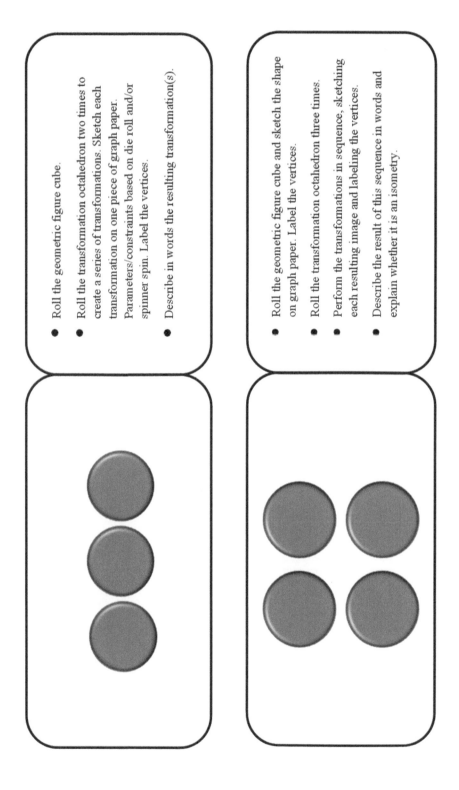

Cubing and Think Dots

- Roll the geometric figure cube, sketch the shape on graph paper and label the vertices.
- Roll the number cube to fill in the coordinates of the translation. (__, __)
- Sketch and give the coordinates of the pre-image.
- Repeat four times using the following signs for each translation:
  (+__, +__); (+__, -__); (-__, -__); (-__, +__)

- Roll the geometric figure cube, sketch the shape on graph paper and label the vertices.
- This is the image after being rotated about the origin clockwise. Spin the spinner to determine the degrees of rotation about the origin.
- Sketch and give the coordinates of the pre-image.
- Repeat the process, but this time rotate counterclockwise about the origin.

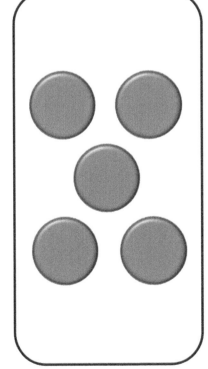
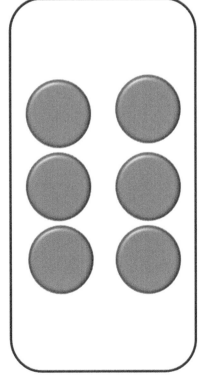

Authentic Opportunities for Writing about Math in High School

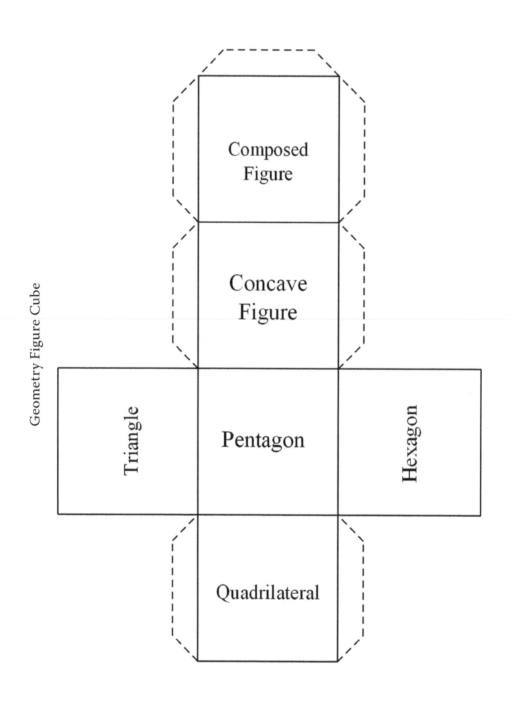

Transformation Octahedron

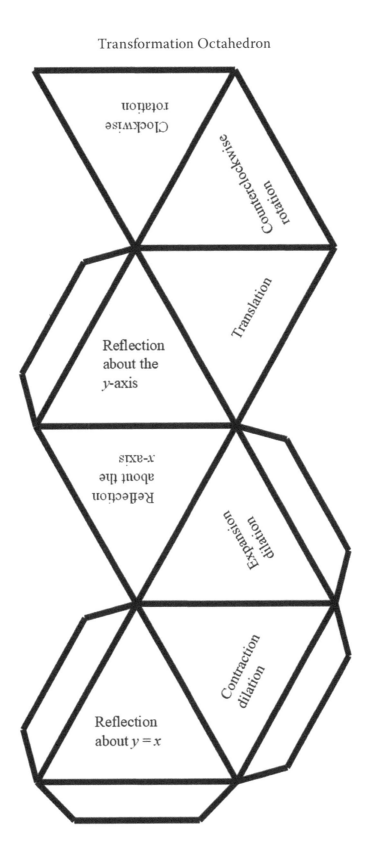

Authentic Opportunities for Writing about Math in High School

Rotation Degree Spinner

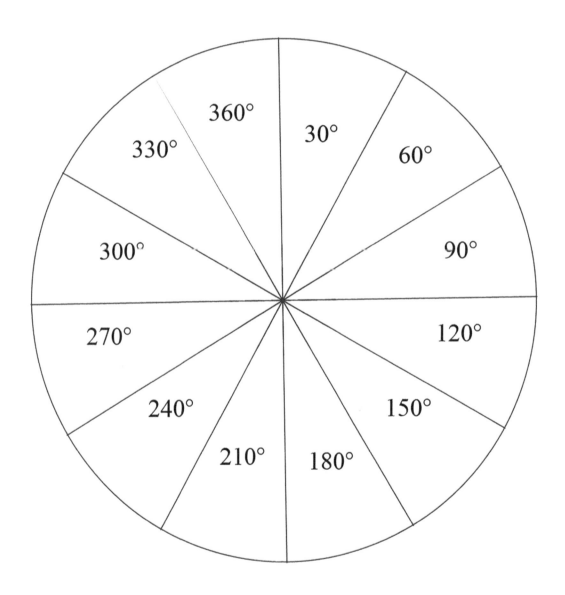

CHAPTER 10

RAFT

The RAFT writing strategy allows students to utilize their creativity within a mathematical context. Students are given a situation, along with a structure, which sets the parameters for the piece. If using as a formative assessment, the topic can be kept vague, as the bold text indicates, to determine whether students have internalized the key understanding(s) of the topic. If the intention is to ensure specific mathematical points are included in the piece, clearly state that expectation within the topic guidelines as the additional text demonstrates.

Role	Audience	Format	Topic
Expression	Equation	Partnership agreement	**Our collaboration**: Discuss how expressions and equations work together.
Circumference of a circle	Area of a circle	Invitation to a family reunion	**How we are related**: Explain the relationship between the circumference and area of a circle.

Authentic Opportunities for Writing about Math in High School

Role	Audience	Format	Topic
System of equations	Jury	Instructions	**Finding solutions:** Discuss how to determine if there is a solution and, if so, how many.
Function	Non-function	Breakup letter	**Irreconcilable differences:** Explain the difference between a function and a non-function.
$\lvert\ \rvert$ (absolute value symbol)	Variable	Complaint	**Mixed signals:** Extend the application of the absolute value symbol from numbers to variables, such as $-\lvert -x \rvert$.
Parallel lines	Transversal	Last will and testament	**Division of assets** Explain the angles formed by cutting parallel lines with a transversal.
Dilation	Translation, rotation, and reflection	Panel discussion	**Request to participate:** Highlight the significant role each transformation, or composition of transformations, plays in advertising. (Or another field, as fits your students' interests.)
Quadratic functions	i	Thank you note	**You are appreciated:** Share statements that acknowledge why i is needed when working with some quadratic functions.
Horizontal line test	Vertical line test	Social media post	**Don't forget about me:** Justify why the vertical line test is not enough when describing functions.

RAFT

Role	Audience	Format	Topic
Interval notation	Function's graph	Text message	**Setting boundaries:** Describe why precise mathematical notation is vital when graphing a function and/or identifying the domain and range.
Algebra, geometry, trig, etc.	Public	Public service announcement	**Why I am important:** Discuss why some working knowledge of my subject is important in today's world.
Student choice	Student choice	Student choice	Student choice

Here is an example of one way that you can "RAFT-ize" a common topic. The set-up below is for a typical coordinate proof for a geometric figure, in this case, a square. This is easily adaptable for other shapes, and even offers students a chance to get creative in writing their own scenario, with illustrations.

Scenario: Quadrilateral SQUR has been seen trying to enter Non-Reg Café for polygons. He is being charged with trespassing. You, as law clerk, have been tasked with writing a legal brief to present to the Grand Jury. The description of the accused has been given as: S(3,3), Q(7,4), U(8,8), R(4,7).

Role	What is the writer's role?	Law clerk
Audience	Who will be reading the piece?	Grand jury
Format	What is the best way to present the information?	Legal brief
Topic	Who or what is the subject?	The evidence of being a square

	Legal brief Submitted by law clerk:
Case name	
Parties involved	
Current state of litigation	
Legal issue	
Relevant facts	
Rule of law applied	
Your argument	
Conclusion	

Facilitation notes: Typically, the components of a legal brief include each of the components listed in the table. For example, the students can be given the coordinates for the four vertices of a square. Sample answers might include the following

Case name: Math University vs Quad SQUR

Parties involved: Segments SQ, QU, UR, and RS, and angles S, Q, U, and R

Current stage of litigation: SQUR plotted on a Cartesian Coordinate System – you can provide a basic one, such as we did, or students can use graph paper.

Legal issue: Is quad SQUR really a square?

Relevant facts: Students should show four congruent sides (using distance formula) and four right angles (adjacent sides perpendicular).

Rule of law: Could be the definition of a square, or related theorem that has already been discussed.

Your argument: Here is where students do the math.

Conclusion: Student's statement of findings.

Additional Format Ideas

Speech	Journal Entry	Script	Song
Brochure	Trading Card	Story Board	Advice Column
Cartoon	Petition	Letter (apology, persuasive, thank you, complaint)	Commercial
Advertisement	Recipe	Campaign Speech	Wanted Poster
Itinerary	Personal Ad	Nursery Rhyme/ Riddle	Infographic
Editorial	Eulogy	Current Event	Debate

Additional Samples

Role: One–half of a Conjugate Pair
Audience: The other half of a Conjugate Pair
Format: Text for a meet up
Topic: What to do when we get together.

Role: Standard form of a quadratic equation
Audience: Binomial Theorem
Format: Proof
Topic: We are just alike.

Role: Complex Fraction
Audience: Algebraic Expressions
Format: Instructions
Topic: Let's simplify!

Role: Real Numbers
Audience: Polynomial Expressions
Format: Review
Topic: Are you closed?

CHAPTER 11

Question Quilt

The question quilt can be a strategy for differentiation that allows for student agency as they read and think about questions and statements relating to a topic of study. They then choose a few of the boxes to think about further. Students decide if they agree or disagree with the statement(s) and/or answer the question(s) and write justification supporting their responses. Questions and statements can be framed to accommodate a variety of levels of learners.

Another option is to give students the question quilt as you are beginning a topic and they can discuss the questions as they progress through the development of the topic.

Sample directions: Choose at least three questions or statements from the question quilt. Answer the question or decide if you agree or disagree with the statement. Justify your responses fully.

Question Quilt

Question Quilt
Functions

- A mathematical function represents a real-world context.

- All functions have solutions.

- Discuss what it means to be a solution to a function and how it compares to a solution for an inequality.

- Choose a representation for a function that you think is the most beneficial and explain why.

- How is it possible for two or more functions that are equivalent to have different algebraic/symbolic forms?

- A function is said to be decreasing on the interval (a, b) if f(x1) < f(x2). A function has a relative minimum when the function changes from decreasing to increasing at a point.

- Functions can be represented in algebraic symbols, contextual situations, graphs, verbal descriptions, tables, sets, and diagrams.

- Operating with functions follows the same rules as operating with real numbers.

- Zeros, roots, and solutions are different.

- Rosa says, "Every function has an inverse." –

Authentic Opportunities for Writing about Math in High School

Question Quilt
Similarity, Right Triangles, and Trigonometry

- Compare and contrast the AA and the HL theorems for triangles.

- What does it mean to "solve a right triangle?"

- How do right triangles relate to the concept of similar triangles?

- Investigate a real-world scenario and determine which trigonometric ratio (sine, cosine, tangent) would be most appropriate to solve the problem.

- Analyze a real-life scenario involving similar triangles and apply the properties of similarity to solve a problem.

- How can congruence and similarity criteria for triangles when proving relationships in geometric figures? Use a specific example.

- Compare and contrast similarity and congruence between geometric figures.

- When solving problems involving right triangles, a more efficient strategy would be to utilize special right triangle relationships.

- $\tan \theta = \dfrac{adjacent}{opposite}$

- How do you explain the practical significance of knowing the relationship between sine and cosine of complementary angles to a peer who has been absent?

CHAPTER 12

Always, Sometimes, and Never

Using always, sometimes, and never questions provides the opportunity for students to investigate a statement and determine whether it is true all the time, some of the time, or never. Typically, always, sometimes, and never have been associated with geometry, but as you can see from the following pages, these are as easily adapted to other topics. They are rich tasks because they engage students in a higher level of reasoning and communication than typical questions might. This activity allows teachers a look into the thinking of the students.

Students may not simply write A, S, or N. They must validate and support their choice. If they determine the statement is sometimes, students should be required to offer a case for when the statement does hold (an example) as well as when it does not (a counter example). Many true and false questions can easily be adapted for always, sometimes, and never statements.

Sample Statements

Topic 1: Number and Quantity

1. Whole numbers are rational.
2. Rational numbers are natural.
3. Integers are whole numbers.
4. A real number is an irrational number.
5. The difference in two whole numbers is a whole number.
6. The product of two integers is a natural number.
7. Fractions are rational numbers.
8. The sum of a rational number and an irrational number is irrational.
9. The number *e* is rational.
10. Inverses operate to give identities.
11. A square number has an even number of factors.
12. $a^b = b^a, a \neq b$
13. Quantities raised to even powers are even.
14. Polynomial long division can be used if the divisor has more than two terms.
15. For any matrix, A_{mxn} for $m \neq n$, A^2 is defined.
16. $x^2 + 25 = (x+5i)(x-5i)$.
17. Matrices are square arrays.
18. Operations with matrices parallel operations with real numbers.
19. The magnitude of a sum of two vectors is the sum of the magnitudes.
20. Expressions with positive exponents can be rewritten with negative exponents only.
21. Expression with negative exponents can be rewritten with positive exponents only.
22. A complex number of the form $a+bi$ will have its corresponding vector lying on the y-axis, provided that $a \neq 0$.
23. You can change the base to base-10 or base-*e* for logarithmic problems?

Topic 2: Algebraic Reasoning

Basic Algebra Review

1. Some equations have no solution.
2. In a linear relationship, if you double a domain value, the corresponding range value is also doubled.
3. $5x > 5+x$

Always, Sometimes, and Never

4. $|2x+3| < 0$
5. $x \leq 7$ is equivalent to $x = 7$
6. The conditionals, "and" and "or", represent the same constraint when working with inequalities.
7. The properties of equality apply in the same way when working with inequalities.
8. A solution represents those values that make a statement true.
9. When you square each side of an equation, the resulting equation is equivalent to the original.
10. Quadratic equations can be solved using the quadratic formula.
11. A function crosses its horizontal asymptote.
12. A function cannot be defined on its vertical asymptotes.

Algebraic Functions

1. Relations are functions.
2. Functions can be evaluated.
3. Input, domain, and x-value are equivalent concepts.
4. Functions are linear.
5. The rate of change does not reveal the relationship between the two attributes defined by the function rule.
6. Functions are horizontal lines.
7. A system of equations consists of two equations.
8. You can determine if a system of equations has a solution through graphing.
9. Systems of equations can be solved using substitution and/or elimination.
10. A system of equations can have multiple solutions.
11. If a solution to a linear system is an ordered pair, the lines have different slopes.
12. The points of intersection that make up the solution to a linear system satisfy both equations simultaneously.
13. Algebraic functions behave similarly when transformed in the coordinate plane.
14. Algebraic functions are functions where one quantity changes at a constant rate per unit interval relative to another.
15. Arithmetic sequences are written recursively and geometric sequences are written with an explicit formula.
16. Function composition is commutative.
17. Function composition is associative, given that the functions are not equal.

18. There is a limit to the number of times you can take a function of a function (of a function ...) with a composite function.
19. The order of the functions does not matter when working with composite functions.
20. The domain of $((x))$ is the same as the domain of $(f(x))$.
21. A function crosses its horizontal asymptote.
22. A function cannot be defined on its vertical asymptotes.

Transcendental Functions

1. Transcendental functions are algebraic functions.
2. The exponential function with base b is defined by $f(x) = b^x$, where $b < 0$ and $b \neq 1$. $f(x) = b^x$, where $b < 0$ and $b \neq 1$.
3. In an exponential function, whenever the input is increased by 1 unit, the output is also increased by the same amount.
4. Exponential functions are characterized by a rate of change that is proportional to the value of the function.
5. The graph of an exponential function approaches but does not touch the x-axis.
6. Exponential functions and logarithmic functions are inverse functions.
7. The domain of a logarithmic function of the form $f(x) = \log_b x$ is the set of all Real Numbers.
8. Trigonometric functions can be undefined for some values of their domain.
9. $Cos(x)$ is an even function.
10. $Tan(x)$ is an even function.
11. In a trigonometric function, the smallest positive value of p for which $f(t+p) = f(t)$ is the period of the function.
12. When graphing a cosecant function, it is helpful to begin by graphing the corresponding cosine function.

Topic 3: Geometric Reasoning/Measurement and Units

Subtopic Specific

Transformations and Congruence

1. If you perform two reflections, it is the same as a translation.
2. Rotations are isometries.
3. The dilation factor changes the perimeter but not the area of a figure.
4. After a dilation, corresponding line segments in an image and its pre-image are parallel.

5. You can change a figure's position without changing its size and shape.
6. The product of two rotations, given different centers, is another rotation.
7. Rigid transformations preserve distance and angle measure.
8. When point P is moved back onto itself under a rigid motion M, P is considered a fixed-point rotation of the rigid motion M.
9. A composition of isometries is an isometry.
10. The result of applying a clockwise rotation followed by a counterclockwise rotation with the same amount of rotation and with the same center of rotation is an identity motion.
11. Composition of transformations is commutative.
12. Two figures are congruent when there is a sequence of transformations that match one figure with the other.
13. Rigid transformations are functions.
14. A curve is point symmetric is it coincides with itself when it is rotated 180° about a point.
15. AA is a triangle congruence relationship.
16. SSA is a triangle congruence relationship.
17. In $\triangle ABC$ and $\triangle XYZ$, if $\overline{AB} \cong \overline{XY}, \angle B \cong \angle Y,$ and $\angle C \cong \angle Z,$ then the triangles must be congruent.
18. The incenter of a triangle is also the center of mass.

Similarity

1. Two triangles are similar.
2. Similar triangles are also congruent.
3. AA is a triangle similarity relationship.
4. The ratio of corresponding sides in similar polygons is constant.
5. Right triangles are similar.
6. If the two vertex angles in two triangles are congruent, the triangles are similar.
7. When working with shadow measurements, using similar triangles is helpful.
8. All circles are similar.
9. Similar polygons have the same area.
10. If the sides of a square are made one-fourth the original length, the area of the new square is one-half the area of the original square.
11. If a line passes through two sides of a triangle, dividing them proportionally, then the line is parallel to the third side.
12. If two triangles are similar, then the corresponding altitudes are congruent to the corresponding sides.
13. If the edge lengths of two similar prisms are in the ratio $\frac{a}{b}$, then the volumes of the prisms are in the ratio of $\frac{a^3}{b^3}$.

Right Triangles and Trigonometry

1. In an isosceles right triangle, the hypotenuse is twice as long as a side.
2. {8, 24, 25} represent the lengths of the sides of a right triangle.
3. The altitude to the hypotenuse of a right triangle forms two triangles that are similar to each other and to the original triangle.
4. Trigonometric functions are periodic functions.
5. If a square and a circle have the same perimeter/circumference, the circle has the smaller area.
6. Values of trigonometric functions are determined for 45°, 45°, 90° and 30°, 60°, 90° triangles.
7. The sine of an acute angles is equal to the cosine of its complement.
8. Standard rotations around the unit circle are performed in a clockwise direction.
9. The Law of Cosines applies to all triangles.
10. For any angle A, $sin^2 A + cos^2 A = 1$.

Circles

1. Any three noncollinear points lie on a circle.
2. The circumference of a circle is irrational.
3. Two circles that are concentric have exactly one common tangent.
4. If a radius intersects a chord in a circle, the point of intersection is the midpoint of the chord.
5. Circles that are tangent have congruent radii.
6. If two chords of a circle are congruent, then the chords are perpendicular.
7. There will always be a dilation that makes two circles have the same radii.
8. If two circles have radii in the ratio $\frac{m}{n}$, then their areas are in the ratio $\frac{m}{n}$.
9. A circle with the equation $(x+6)^2 + (y-7)^2 = 64$ does not include points in Quadrant IV.
10. A circle is an ellipse.

Proof, Logic, and Reasoning

1. Inductive reasoning leads to valid conclusions.
2. Axioms are used in geometry.
3. A statement is a declarative sentence having truth value.
4. Questions and commands are statements.

5. It is impossible to think of a situation in which a statement and its negation will have the same truth value.
6. Any two statements p and q are logically equivalent.
7. When writing a geometric proof, the supplementary angle relationship could be used alone to justify that two angles are congruent.
8. Triangle TRI has vertices T (0, 0), R (0,4), and I (3,2). The triangle can be classified as isosceles.
9. In a coordinate plane, the locus of points 5 units from the x-axis is the lines x = 5 and x = −5.

2-D and 3-D Geometric Measurement

1. Slices of a cone are circles.
2. 3-D figures are polyhedrons.
3. Slices of a sphere passing through the center of the sphere are congruent.
4. The sum of the measures of the interior angles of a hexagon equal the sum of the measures of the exterior angles.
5. Figures that have equal areas are congruent.
6. Having rotational symmetry is the same as having both horizontal and vertical symmetry.
7. The surface area of a sphere is equal to the volume of the same sphere.
8. If an isosceles triangle is rotated about the altitude from its vertex, the resulting figure is a pyramid.
9. A cross section of a cube can be an octagon.
10. When two solids have identical heights and cross-sections, the volumes of the two figures are equal.

Vectors

1. A vector is defined by its magnitude, or the length of the line, and its direction, indicated by an arrowhead at the terminal point.
2. Vectors are dependent on the coordinate system.
3. Two vectors are parallel if they have the same magnitude and direction.
4. The sum of two vectors is called the resultant.
5. The sum of two vectors is the sum of the magnitudes.
6. We add two vectors together by positioning the head of the first vector at the head of the second vector.
7. The actual location of a vector within the 2D plane is unimportant because a vector represents a translation rather than an object with a fixed position.
8. Parallel vectors can be written as nonzero scalar multiples of each other.
9. Subtracting a vector is the same as adding its negative.
10. Scalar multiplication has no effect on the direction.

Topic 4: Data Analysis, Probability, and Statistics

Data Analysis and Statistics

1. The standard deviation is the average amount of variability in your dataset.
2. A high standard deviation means that values are generally far from the mean, while a low standard deviation indicates that values are clustered close to the mean.
3. Data with a larger range has a higher standard deviation.
4. Normal distribution is a probability distribution that is asymmetric about the mean.
5. In graphical form, the normal distribution appears as a "bell curve."
6. The two main parameters of a (normal) distribution are the mean and the median.
7. Parameters determine the shape and probabilities of the distribution.
8. Causation does not imply correlation.
9. Correlation implies causation.
10. Measures of central tendency can be determined for a bivariate data set.
11. A t test can be used when comparing the means of multiple groups. **S**
12. Linear regression can be used to: (1) determine the strength of predictors, (2) forecast an effect, and (3) for trend forecasting.
13. Descriptive inference is the process of analyzing the result and making conclusions from data subject to random variation.
14. The null hypothesis means there is no connection among groups or no association between two measured events.
15. Discrete data involves numbers.
16. Skewness describes how much statistical data distribution is asymmetrical from the normal distribution.
17. The least square method is a way of finding the line of best fit.
18. Least squares regression is used to predict the behavior of independent variables.
19. The lower the interquartile range, the more spread out the data points.
20. An outlier is an observation that lies an abnormal distance from other values in a random sample from a population.
21. If the points are further from the mean, there is a higher deviation within the data.

Always, Sometimes, and Never

Probability

1. The probability of an event is a non-negative real number and equal to or smaller than 1.
2. The probability of an event A is the number of outcomes in A divided by the number of outcomes in the sample space.
3. The probability of two events occurring is always greater than 0.
4. The conjunction "and", when used in probability, means to add.
5. The probability of a compound event is the percentage of outcomes in the sample space for which the compound event occurs.
6. Conditional probability means the problem presents a condition where calculating the probability is possible.
7. Geometric probability is a volume model.
8. $P(A|B) = P(A)$.
9. $P(A,B) = P(A)P(B)$.
10. $P(A|B) = \dfrac{P(A,B)}{P(B)} \equiv \dfrac{P(A \text{ and } B)}{P(B)} \equiv \dfrac{P(A \cap B)}{P(B)}$.
11. Permutation problems involve situations in which order matters.
12. Combination problems involve situations in which order matters.
13. If you have a problem where you can repeat objects, then you must use the Fundamental Counting Principle, you can't use permutations or combinations.
14. A permutation is an ordered Combination.
15. The **factorial function** (!) means to multiply a series of descending natural numbers.
16. The probability of it snowing tomorrow is .5 because it will either snow or not snow.
17. Expected value is a weighted average.
18. Normal distribution is a probability distribution that is asymmetric about the mean.
19. In graphical form, the normal distribution appears as a "bell curve."
20. The two main parameters of a (normal) distribution are the mean and the median.
21. Parameters determine the shape and probabilities of the distribution.

Sample Statements with Answers

Topic 1: Number and Quantity

1. Whole numbers are rational. **(A)**
2. Rational numbers are natural. **(A)**
3. Integers are whole numbers. **(S)**
4. A real number is an irrational number. **(S)**
5. The difference in two whole numbers is a whole number. **(S)**
6. The product of two integers is a natural number. **(S)**
7. Fractions are rational numbers. **(S)**
8. The sum of a rational number and an irrational number is irrational. **(A)**
9. The number e is rational. **(N)**
10. Inverses operate to give identities. **(A)**
11. A square number has an even number of factors. **(N)** **(Note: It is always an odd number of factors, since you only count the square root once in the list of factors.)**
12. $a^b = b^a, a \neq b$ **(S)** (Note: Students can investigate this using simple substitution as needed. $2^4 = 4^2$)
13. Quantities raised to even powers are even. **(S)**
14. Polynomial long division can be used if the divisor has more than two terms. **(A)**
15. For any matrix, A_{mxn} for $m \neq n$, A^2 is defined. **(N)**
16. $x^2 + 25 = (x+5i)(x-5i)$. **(A)**
17. Matrices are square arrays. **(S)**
18. Operations with matrices parallels operations with real numbers. **(S)**
19. The magnitude of a sum of two vectors is the sum of the magnitudes. **(S)**
20. Expressions with positive exponents can be rewritten with negative exponents only. **(A)**
21. Expression with negative exponents can be rewritten with positive exponents only. **(A)**
22. A complex number of the form $a+bi$ will have its corresponding vector lying on the y-axis, provided that $a \neq 0$. **(N)**
23. You can change the base to base-10 or base-e for logarithmic problems? **(S)**

Always, Sometimes, and Never

Topic 2: Algebraic Reasoning

Basic Algebra Review

1. Some equations have no solution. **(A)**
2. In a linear relationship, if you double a domain value, the corresponding range value is also doubled. **(A)**
3. $5x > 5+x$ **(S)**
4. $|2x+3| < 0$ **(N)**
5. $x \leq 7$ is equivalent to $x = 7$ **(N)**
6. The conditionals, "and" and "or", represent the same constraint when working with inequalities. **(N)**
7. The properties of equality apply in the same way when working with inequalities. **(S)**
8. A solution represents those values that make a statement true. **(A)**
9. When you square each side of an equation, the resulting equation is equivalent to the original. **(A)**
10. Quadratic equations can be solved using the quadratic formula. **(A)**
11. A function crosses its horizontal asymptote. **(N)**
12. A function cannot be defined on its vertical asymptotes. **(A)**

Algebraic Functions

1. Relations are functions. **(S)**
2. Functions can be evaluated. **(A)**
3. Input, domain, and x-value are equivalent concepts. **(A)**
4. Functions are linear. **(S)**
5. The rate of change does not reveal the relationship between the two attributes defined by the function rule. **(S)**
6. Functions are horizontal lines. **(S)**
7. A system of equations consists of two equations. **(S)**
8. You can determine if a system of equations has a solution through graphing. **(S)**
9. Systems of equations can be solved using substitution and/or elimination. **(S)**
10. A system of equations can have multiple solutions. **(S)**
11. If a solution to a linear system is an ordered pair, the lines have different slopes. **(A)**
12. The points of intersection that make up the solution to a linear system satisfy both equations simultaneously. **(A)**

13. Algebraic functions behave similarly when transformed in the coordinate plane. **(A)** (**Note: horizontal and vertical shifts as well as compressing and stretching and end behaviors work generally the same way.**)
14. Algebraic functions are functions where one quantity changes at a constant rate per unit interval relative to another. **(A)**
15. Arithmetic sequences are written recursively and geometric sequences are written with an explicit formula. **(S)**
16. Function composition is commutative. **(S)**
17. Function composition is associative, given that the functions are not equal. **(N)**
18. There is a limit to the number of times you can take a function of a function (of a function …) with a composite function. **(N)**
19. The order of the functions does not matter when working with composite functions. **(N)**
20. The domain of $(f(x))$ is the same as the domain of $(f(x))$. **(S)**
21. A function crosses its horizontal asymptote. **(N)**
22. A function cannot be defined on its vertical asymptotes. **(A)**

Transcendental Functions

1. Transcendental functions are algebraic functions. **(N)**
2. The exponential function with base b is defined by $f(x) = b^x$, where $b < 0$ and $b \neq 1$. $f(x) = b^x$, where $b < 0$ and $b \neq 1$. **(N)**
3. In an exponential function, whenever the input is increased by 1 unit, the output is also increased by the same amount. **(N)**
4. Exponential functions are characterized by a rate of change that is proportional to the value of the function. **(A)**
5. The graph of an exponential function approaches but does not touch the x-axis. **(S)** (**the graph of the transformation $f(x) = 2^x - 1$ does cross the x-axis at the origin.**)
6. Exponential functions and logarithmic functions are inverse functions. **(A)**
7. The domain of a logarithmic function of the form $f(x) = \log_b x$ is the set of all Real Numbers. **(N)** (**Can only be positive real numbers.**)
8. Trigonometric functions can be undefined for some values of their domain. **(S)**
9. $\cos(x)$ is an even function. **(A)**
10. $\tan(x)$ is an even function. **(N)**

11. In a trigonometric function, the smallest positive value of p for which $f(t+p) = f(t)$ is the period of the function. **(A)**
12. When graphing a cosecant function, it is helpful to begin by graphing the corresponding cosine function. **(N)**

Topic 3: Geometric Reasoning/Measurement and Units

Subtopic Specific

Transformations and Congruence:
1. If you perform two reflections, it is the same as a translation. **(S) (If the reflections are across two parallel lines.)**
2. Rotations are isometries. **(A)**
3. The dilation factor changes the perimeter but not the area of a figure. **(N)**
4. After a dilation, corresponding line segments in an image and its pre-image are parallel. **(S)**
5. You can change a figure's position without changing its size and shape. **(A)**
6. The product of two rotations, given different centers, is another rotation. **(S) (Could be a translation – two half-turns)**
7. Rigid transformations preserve distance and angle measure. **(A)**
8. When point P is moved back onto itself under a rigid motion M, P is considered a fixed-point rotation of the rigid motion M. **(A)**
9. A composition of isometries is an isometry. **(A)**
10. The result of applying a clockwise rotation followed by a counterclockwise rotation with the same amount of rotation and with the same center of rotation is an identity motion. **(A)**
11. Composition of transformations is commutative. **(S)**
12. Two figures are congruent when there is a sequence of transformations that match one figure with the other. **(A)**
13. Rigid transformations are functions. **(A)**
14. A curve is point symmetric is it coincides with itself when it is rotated 180° about a point. **(A)**
15. AA is a triangle congruence relationship. **(N)**
16. SSA is a triangle congruence relationship. **(N)**
17. In $\triangle ABC$ and $\triangle XYZ$, if $\overline{AB} \cong \overline{XY}, \angle B \cong \angle Y,$ and $\angle C \cong \angle Z$, then the triangles must be congruent. **(A)**
18. The incenter of a triangle is also the center of mass. **(S) (In an equilateral triangle, all four points of concurrency are the same, otherwise, the center of mass is the centroid.)**

Similarity

1. Two triangles are similar. **(S)**
2. Similar triangles are also congruent. **(S)**
3. AA is a triangle similarity relationship. **(A)**
4. The ratio of corresponding sides in similar polygons is constant. **(A)**
5. Right triangles are similar. **(S)**
6. If the two vertex angles in two triangles are congruent, the triangles are similar. **(A)**
7. When working with shadow measurements, using similar triangles is helpful. **(S)**
8. All circles are similar. **(A)**
9. Similar polygons have the same area. **(S)**
10. If the sides of a square are made one-fourth the original length, the area of the new square is one-half the area of the original square. **(N)**
11. If a line passes through two sides of a triangle, dividing them proportionally, then the line is parallel to the third side. **(A)**
12. If two triangles are similar, then the corresponding altitudes are congruent to the corresponding sides. **(N)**
13. If the edge lengths of two similar prisms are in the ratio $\frac{a}{b}$, then the volumes of the prisms are in the ratio of $\frac{a^3}{b^3}$. **(C)**

Right Triangles and Trigonometry

1. In an isosceles right triangle, the hypotenuse is twice as long as a side. **(N)**
2. {8, 24, 25} represent the lengths of the sides of a right triangle. **(N)**
3. The altitude to the hypotenuse of a right triangle forms two triangles that are similar to each other and to the original triangle. **(A)**
4. Trigonometric functions are periodic functions. **(A)**
5. If a square and a circle have the same perimeter/circumference, the circle has the smaller area. **(N)**
6. Values of trigonometric functions are determined for 45°, 45°, 90° and 30°, 60°, 90° triangles. **(S)**
7. The sine of an acute angles is equal to the cosine of its complement. **(A)**
8. Standard rotations around the unit circle are performed in a clockwise direction. **(N)**
9. The Law of Cosines applies to all triangles. **(S)**
10. For any angle A, $sin^2 A + cos^2 A = 1$. **(A)**

Always, Sometimes, and Never

Circles

1. Any three noncollinear points lie on a circle. **(A)**
2. The circumference of a circle is irrational. **(S)**
3. Two circles that are concentric have exactly one common tangent. **(N)**
4. If a radius intersects a chord in a circle, the point of intersection is the midpoint of the chord. **(S)**
5. Circles that are tangent have congruent radii. **(S)**
6. If two chords of a circle are congruent, then the chords are perpendicular. **(S)**
7. There will always be a dilation that makes two circles have the same radii. **(A)**
8. If two circles have radii in the ratio $\frac{m}{n}$, then their areas are in the ratio $\frac{m}{n}$. **(N)**
9. A circle with the equation $(x+6)^2 + (y-7)^2 = 64$ does not include points in Quadrant IV. **(A)**
10. A circle is an ellipse. **(A)**

Proof, Logic, and Reasoning

1. Inductive reasoning leads to valid conclusions. **(S)**
2. Axioms are used in geometry. **(S)**
3. A statement is a declarative sentence having truth value. **(A)**
4. Questions and commands are statements. **(N)**
5. It is impossible to think of a situation in which a statement and its negation will have the same truth value. **(A)**
6. Any two statements p and q are logically equivalent. **(S)**
7. When writing a geometric proof, the supplementary angle relationship could be used alone to justify that two angles are congruent. **(S)**
8. Triangle TRI has vertices T (0, 0), R (0,4), and I (3,2). The triangle can be classified as isosceles. **(A)**
9. In a coordinate plane, the locus of points 5 units from the x-axis is the lines x = 5 and x = −5. **(N)**

2-D and 3-D Geometric Measurement

1. Slices of a cone are circles. **(S)**
2. 3-D figures are polyhedrons. **(S)**
3. Slices of a sphere passing through the center of the sphere are congruent. **(A)**

4. The sum of the measures of the interior angles of a hexagon equal the sum of the measures of the exterior angles. **(N)**
5. Figures that have equal areas are congruent. **(S)**
6. Having rotational symmetry is the same as having both horizontal and vertical symmetry. **(S)**
7. The surface area of a sphere is equal to the volume of the same sphere. **(S)**
8. If an isosceles triangle is rotated about the altitude from its vertex, the resulting figure is a pyramid. **(N)**
9. A cross section of a cube can be an octagon. **(N)**
10. When two solids have identical heights and cross-sections, the volumes of the two figures are equal. **(S)**

Vectors

1. A vector is defined by its magnitude, or the length of the line, and its direction, indicated by an arrowhead at the terminal point. **(A)**
2. Vectors are dependent on the coordinate system. **(N)**
3. Two vectors are parallel if they have the same magnitude and direction. **(S)**
4. The sum of two vectors is called the resultant. **(A)**
5. The sum of two vectors is the sum of the magnitudes. **(S)**
6. We add two vectors together by positioning the head of the first vector at the head of the second vector. **(N)**
7. The actual location of a vector within the 2D plane is unimportant because a vector represents a translation rather than an object with a fixed position. **(A)**
8. Parallel vectors can be written as nonzero scalar multiples of each other. **(A)**
9. Subtracting a vector is the same as adding its negative. **(A)**
10. Scalar multiplication has no effect on the direction. **(S)**

Topic 4: Data Analysis, Probability, and Statistics

Data Analysis and Statistics

1. The standard deviation is the average amount of variability in your dataset. **(A)**
2. A high standard deviation means that values are generally far from the mean, while a low standard deviation indicates that values are clustered close to the mean. **(A)**
3. Data with a larger range has a higher standard deviation. **(A)**

Always, Sometimes, and Never

4. Normal distribution is a probability distribution that is asymmetric about the mean. **(N)**
5. In graphical form, the normal distribution appears as a "bell curve." (A)
6. The two main parameters of a (normal) distribution are the mean and the median. **(N)**
7. Parameters determine the shape and probabilities of the distribution. **(A)**
8. Causation does not imply correlation. **(N)**
9. Correlation implies causation. **(S)**
10. Measures of central tendency can be determined for a bivariate data set. **(N)**
11. A t test can be used when comparing the means of multiple groups. **(S)**
12. Linear regression can be used to: (1) determine the strength of predictors, (2) forecast an effect, and (3) for trend forecasting. **(A)**
13. Descriptive inference is the process of analyzing the result and making conclusions from data subject to random variation. **(N)**
14. The null hypothesis means there is no connection among groups or no association between two measured events. **(A)**
15. Discrete data involves numbers. **(S)**
16. Skewness describes how much statistical data distribution is asymmetrical from the normal distribution. **(A)**
17. The least square method is a way of finding the line of best fit. **(S)**
18. Least squares regression is used to predict the behavior of independent variables. **(N)**
19. The lower the interquartile range, the more spread out the data points. **(N)**
20. An outlier is an observation that lies an abnormal distance from other values in a random sample from a population. **(S) (If "abnormal" is defined to be 1.5 IQR)**
21. If the points are further from the mean, there is a higher deviation within the data. **(A)**

Probability

1. The probability of an event is a non-negative real number and equal to or smaller than 1. **(A)**
2. The probability of an event A is the number of outcomes in A divided by the number of outcomes in the sample space. **(S) (only true when all the outcomes in the sample space have the same probability of happening)**.

3. The probability of two events occurring is always greater than 0. **(S) (if we take two events with a probability of zero, then the probability of these two events occurring is zero).**
4. The conjunction "and", when used in probability, means to add. **(N)**
5. The probability of a compound event is the percentage of outcomes in the sample space for which the compound event occurs. **(A)**
6. Conditional probability means the problem presents a condition where calculating the probability is possible. **(N)**
7. Geometric probability is a volume model. **(N)**
8. $P(A|B) = P(A)$. **(S) (if A and B are independent)**
9. $P(A,B) = P(A)P(B)$. **(S) (if A and B are independent)**
10. $P(A|B) = \dfrac{P(A,B)}{P(B)} \equiv \dfrac{P(A \text{ and } B)}{P(B)} \equiv \dfrac{P(A \cap B)}{P(B)}$. **(A)**
11. Permutation problems involve situations in which order matters. **(A)**
12. Combination problems involve situations in which order matters. **(N)**
13. If you have a problem where you can repeat objects, then you must use the Fundamental Counting Principle, you can't use permutations or combinations. **(A)**
14. A permutation is an ordered Combination. **(A)**
15. The factorial **function** (!) means to multiply a series of descending natural numbers. **(A)**
16. The probability of it snowing tomorrow is .5 because it will either snow or not snow. **S**
17. Expected value is a weighted average. **(A)**
18. Normal distribution is a probability distribution that is asymmetric about the mean. **(N)**
19. In graphical form, the normal distribution appears as a "bell curve." **(A)**
20. The two main parameters of a (normal) distribution are the mean and the median. **(N)**
21. Parameters determine the shape and probabilities of the distribution. **(A)**
22. Scoring a sum of "3" with two number cubes is twice as likely as scoring a sum of "2." **(A)**
23. The experimental probability of getting exactly two heads tossing four two-sided coins is one-half. **(S)**
24. The probability of an event happening can be greater than 1. **(N)**
25. The theoretical probability of an event happening is the same as the experimental probability of the same event happening. **(S)**
26. The conjunction "and", when used in probability, means to add. **(N)**

27. The probability of a compound event is the percentage of outcomes in the sample space for which the compound event occurs. **(A)**
28. Tree diagrams and tables are the only ways to represent sample spaces for compound events. **(N)**
29. The probability of it snowing tomorrow is .5 because it will either snow or not snow. **(S)**
30. A probability represented by a percentage is different from a probability represented by the equivalent fraction. **(N)**
31. The basic definition for probability can be represented by the ratio $\frac{n(E)}{n(S)}$, where n is the number of E (events) divided by the number in the S (sample space). **(A)**

PART THREE

Planning and Implementation

CHAPTER 13

Crosswalk

The following crosswalk is included to support instructional planning. Resources can be quickly identified based upon the mathematical topic as well as the type of writing and/or the strategy example given. The crosswalk identifies the mathematical topics that are included in the given examples referenced in each of the 11 writing strategies shared in the previous chapter.

Authentic Opportunities for Writing about Math in High School

Crosswalk of Topics and Writing Strategies

Writing Strategy \ Topics	Number and Quantity	Algebraic Reasoning	Geometric Reasoning/ Measurement and Units	Data Analysis, Probability, and Statistics	Universal
Always, Sometimes, and Never	x	x	x	x	
Question Quilts		x	x		
RAFTs	x	x	x		
Cubing/Think Dots		x	x		
Poems	x		x		x
Journal Prompts	x	x	x	x	x
Writing About	x	x	x	x	
Topical Questions	x	x	x	x	
The Answer Is…	x	x	x	x	
Compare/Contrast	x	x	x	x	
Visual Prompts		x	x		x

170

CHAPTER 14

Bringing It All Together

This last section provides a sample anchor task that demonstrates how several of these writing strategies can be authentically integrated into classroom instruction. A lesson plan and facilitation notes are provided.

Farmer Jones

Overview: This task was intentionally chosen because it not only models several opportunities for writing but also demonstrates the efficiency of planning for addressing multiple content and process standards. Planning for simultaneous outcomes allows time for students to dive deep into a single task rather than completing multiple tasks over the same timeframe. It allows students to make connections within the content rather than viewing the concepts as discrete and unrelated. This task is so versatile it can be modified in unlimited ways to meet the needs of both teachers and students.

The Farmer Jones tasks for high school students require students to investigate the relationships between the tangram pieces while working on an open-ended problem. One activity requires students to maximize income and profit while the second asks for a minimization. Developing different options and presenting those options is required. Even though this is a contrived problem, students can make it more authentic by researching the crops and

Authentic Opportunities for Writing about Math in High School

livestock options in their geographic region. The outline below is just one suggestion for how these can be facilitated.

Writing Opportunities

- ❏ Visual prompt (p. 17)
- ❏ Topical questions – "We're Stuck/"We're Done" (p. 13 and p. 39)
- ❏ Problem-solving process (p. 11–12)
- ❏ Reflection (p. 160)

Content Connections: (This task can be used across courses so possible topics may include but are not limited to)

- ❏ Reason quantitatively and use units to solve problems.
- ❏ Geometric reasoning – special right triangles, Pythagorean Theorem
- ❏ Real world context (connections to other disciplines –- financial planning)
- ❏ Problem-solving (standards for mathematical practice)

Routines

The 5 Practices for Orchestrating Productive Mathematics Discussions

- ❏ Anticipating students' solutions to a mathematics task
- ❏ Monitoring students' in-class, "real-time" work on the task
- ❏ Selecting approaches and students to share them
- ❏ Sequencing students' presentations purposefully
- ❏ Connecting students' approaches and the underlying mathematics

Materials Needed

- ❏ Tangrams, two sets per student pair (See **Preparation for Implementation: PLC Work**)
- ❏ Visual of tangram square
- ❏ Tasks handouts – depending on which one is selected.
- ❏ Problem Solving Process Organizer
- ❏ "We're Stuck" and "We're Done" questions
- ❏ QUAD Reflection

Bringing It All Together

Preparation

It is more beneficial to have students cut the tangrams. When students create their own tangram set, it allows them to better observe the relationships between the pieces and offers you the opportunity to monitor/pre-assess their vocabulary and prior geometric work. Directions for paperfolding are found in the **Preparation for Implementation: PLC Work.** Alternatively, students can use the blackline master to make tangram sets.

Complete the task as if you are a student. Anticipate student strategies, solutions, misconceptions, etc. This process informed the development of the "We're Stuck" and "We're Done" questions used to support students. See **Preparation for Implementation: PLC Work** for a more detailed explanation.

Warm up: 10 Minutes

Visual Prompt: Project the tangram master, or an image of the completed tangram square.

Have students sketch the image in the Mathematician's Notebook and respond to the following questions using precise mathematical language:

What shapes can you name?
How are the shapes alike?
How are they different?

Activity 1: 15 Minutes

Give each student a piece of cardstock, construction paper, etc. Together as a class, complete the paper folding and/or the cutting out of the tangram sets. See **Preparation for Implementation: PLC Work** for a more detailed explanation.

Activity 2: 45 Minutes

Introduce the selected task by displaying it on the screen and reading aloud.

Distribute the handout and ask students to begin by doing some independent thinking. Reinforce the expectations by displaying and stating the following:

Think about the problem

Write by doing one of the following:

❏ Write the question that is being asked and/or create an answer statement.

- ❏ Write all facts you know about the problem.
- ❏ Write what additional information is needed or questions you have about the problem.

Providing these questions allows access for all students and eliminates "no response" as an option. NOTE: If students are using the problem-solving process graphic organizer, there is a copy of the organizer for support in the workspace section. Have students partner to share strategies and questions. This should be done intentionally using a collaborative structure and strategy to ensure equity in each pair with both students having a voice. Students should annotate their work as needed during and after the discussions.

Differentiation Strategy: As students are working, allow them to access to the scaffolded "We're Stuck" questions. This can be done by having students get them from a central location (hard copy) or access on a device (electronic copy). These can also be placed on student desks while monitoring and identifying where students may need support. As students complete the task, monitor to make sure they are utilizing the "We're Done" questions.

Monitor student pairs and identify various strategies, misconceptions, "ah-ha" moments, etc. to highlight in the whole class discussion at the conclusion of the activity. Select (and inform) students prior to the discussion that they will be asked to share their thinking. Be specific in telling them what you would like them to share when called upon. This will keep the discussion focused on the main ideas.

Lesson Synthesis: 25 minutes

Conduct a classroom discussion by sequencing the student groups who were selected to share. This should be done in a manner that allows connections to be made between approaches and ideas to reinforce the learning goals set forth in the activity.

Student Reflection: 5 Minutes

Have students complete the QUAD Reflection independently.

- ❏ **Question:** What questions do you still have about the problem?
- ❏ **Understanding:** What do you now understand after working with this problem?
- ❏ **Activate:** How did this problem activate you as a learning resource for a peer or a peer for you?
- ❏ **Discourse:** What mathematical discourse was prompted by working this problem?

Preparation for Implementation: PLC Work

The best way to prepare for the implementation of each task is for teachers to meet to unpack and discuss. There are two tasks available. Farmer Jones 1 is designed for more introductory work while Farmer Jones 2 provides more cognitive demand. Some ideas are outlined here.

1. Have all teachers complete the selected task as if they were students using the handout provided.
2. Discuss strategies and solutions from the group as well as other potential approaches. In addition, answer the questions below.
   a) Where will students demonstrate success with the task?
   b) Where will students struggle with the task? (OMG's – SREB abbreviation)
      ❏ Obstacles – Students have a lack of understanding of which strategies or procedures to apply and how those strategies work.
      ❏ Misconceptions – Students are unaware that the knowledge they have is incorrect.
      ❏ Gaps in Learning – Students lack prerequisite knowledge. (SREB, 2018)
3. Based on possible student struggles with the task, review the "stuck" questions that have been generated for support. Edit or add questions that may be needed. Keep in mind to limit the total number to between three and five so students don't get overwhelmed.
   ❏ How does making the tangram square help in working this problem?
   ❏ What special right triangle relationships do you observe?
   ❏ What is the relationship between each of the pieces of the tangram?
   ❏ What do you need to know to determine the fencing required for each piece of land?
   ❏ What is the relationship between each of the pieces of the tangram and the whole tract of land? How will this help determine where to place the different planting and/or grazing options?
4. Complete the same process with the "done" questions.
   ❏ How can three different geometric shapes of land have the same area?

- How did the yield per acre of each piece of land impact your decision on the plot's use?
- How would the values change if the total area of the land was an acre or more?
- How would you justify the options you determined for each piece of land?
- What specifically in your first layout informed your decisions as you created the second layout? Why? (IMPORTANT – This question is intentionally designed to cause students to think more deeply about crop yields, etc..)

> NOTE: If writing additional questions, keep in mind they…
> - Arise out of students' misconceptions
> - Cause the student to think more deeply about the mathematics
> - Should be answerable by the teacher
> - Should be answerable by more than a yes or a no
> - Can be direct

NOTE: Tangrams can be cut from die cuts and foam, cardstock, construction paper, etc. Alternatively, students can make their own set of tangrams through paper folding and tearing/cutting. This is a good spatial reasoning exercise. Directions are below.

Paper Folding a Tangram

- Square up a piece of 8.5" x 11" paper. Fold the paper so a shorter side lies on tops of (coincides) with one of the longer sides. Fold back and forth, creasing each time, and tear off the rectangle. You should now have a square piece of paper and a rectangle. Keep the square.
- Fold the square along one diagonal and crease to make two congruent right triangles. Fold back and forth, creasing each time, and tear apart the right triangles. Set one aside.
- Take one right triangle and fold in half so you form another set of two congruent right triangles. Fold back and forth, creasing each time, and tear apart the right triangles. Set these aside. They are the first two tangram pieces.
- Take the second large right triangle and position it so the right triangle is at the top and the hypotenuse is the base. Fold the right angle (the square corner) down to the middle of the opposite side (the

hypotenuse). Fold back and forth, creasing each time, and tear apart the small triangle on top from the isosceles trapezoid on the bottom. Set the triangle aside. This is the third tangram piece.

❏ Turn the isosceles trapezoid so the longer base is on the bottom. Fold the left side of the trapezoid over on top of the right side so you have folded it in half along a vertical line of symmetry. Unfold and fold the left bottom corner to the middle fold line so the bottom sides lie on top of each other. Crease well and tear apart the small triangle and the remaining small square on the left of the fold line. These are the fourth and fifth pieces of the tangram.

❏ Take the remaining trapezoid and turn it so the right angles are on the left and the longer base is on the bottom. Take the upper left corner (at the obtuse angle) and fold it down and to the left corner (the lower right angle) so the bottom sides lie on top of each other. Crease well and tear apart the small triangle on the left and the remaining parallelogram. These are the sixth and seventh pieces of the tangram.

Farmer Jones
Tangram Relationship Hints

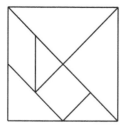

Tangram Square
Relationship Hints

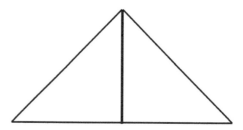

2 large triangles = $\frac{1}{2}$ the tract of land

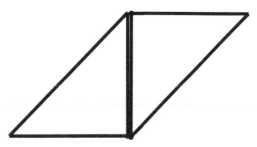

2 small triangles = 1 parallelogram

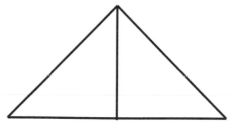

2 medium triangles = 1 large triangle

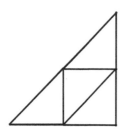

4 small triangles = 1 large triangle

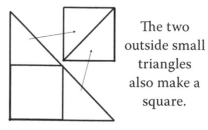

2 squares = 1 large triangle

The two outside small triangles also make a square.

Bringing It All Together

Tangram Master

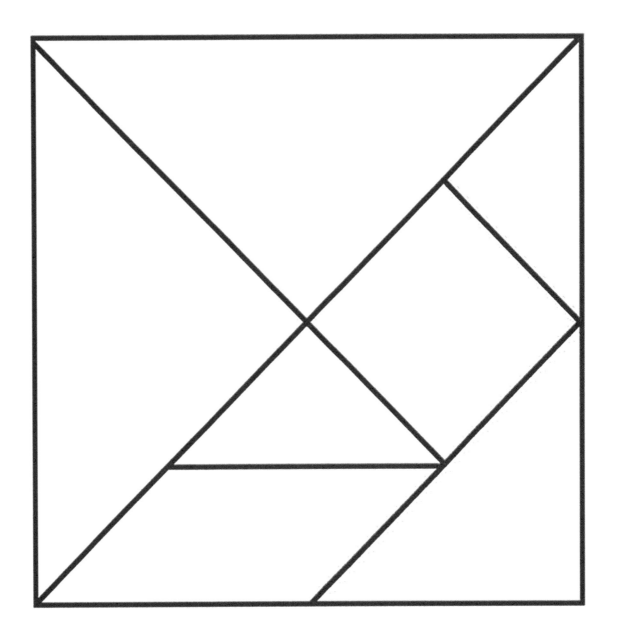

Farmer Jones Task 1

Farmer Jones has a square field, 2000 feet per side, divided into parcels that make a tangram (to please Mrs. Jones). One side of the land is next to a deep river. The cows need access to water. All sides need fencing, with different fencing required depending on what that fence contains. Fencing requirements along with other data is provided in the given table. Determine a planting scheme that minimizes fencing cost. Prepare a presentation that gives Farmer Jones two options. Elaborate on the principles you used to make the planting and livestock choices. Provide estimated incomes from both options.

Item growing	Fencing cost per foot	Yield/acre
Dairy Cows	$85	$500
Cashmere Goats	$95	$800
Corn	$25	$200
Watermelon	$35	$450
Wheat	$15	$1000
Soybeans	$10	$750
String beans	$20	$900

Farmer Jones Task 2

Farmer Jones has a square field, 2000 feet per side, divided into parcels shaped like a tangram (to please Mrs. Jones). Farmer Jones must plant each of four different crops in the different parcels, one crop per parcel. His planting options are provided in the table below. Crops need more expensive fencing when they are subject to predatory animals. Devise a planting scheme for Farmer Jones and estimate his income. List the different ideas to consider to (1) maximize income and (2) maximize profit. Provide Farmer Jones with two options to consider.

Crop	Fertilizer/acre	Fencing/ft	Yield/acre
Tomatoes	$15	$12	$300
Watermelon	$7	$20	$150
Corn	$10	$23	$125
Spaghetti Squash	$7	$10	$350

Problem-Solving Process

The problem is asking me to...	I know...
Answer statement:	
Topic/concept this is related to...	**Strategy for solving...**

Solve (show work here)

This solution means...

Farmer Jones: We're Stuck

1. How does making the tangram square help in working this problem?
2. What special right triangle relationships do you observe?
3. What is the relationship between each of the pieces of the tangram?
4. What do you need to know to determine the fencing required for each piece of land?
5. What is the relationship between each of the pieces of the tangram and the whole tract of land?

How will this help determine where to place the different planting and/or grazing options?

Farmer Jones: We're Done

1. How can three different geometric shapes of land have the same area?
2. How did the yield per acre of each piece of land impact your decision on the plot's use?
3. How would the values change if the total area of the land was an acre or more?
4. How would you justify the options you determined for each piece of land?
5. What specifically in your first layout informed your decisions as you created the second layout? Why?

QUAD Reflection

Q: Question:
What questions do you still have about the task?

U: Understanding:
What do you now understand after working with this task?

A: Activate:
How did this task activate you as a learning resource for a peer or a peer for you?

D: Discourse:
What discourse did working this task prompt?

Afterword

We have always had educators ask us if they could get a list of the questions we used during the day's training, a list of the writing prompts we used, an example of a DI strategy we mentioned, etc. Our questions came from our interactions with the participants; the writing prompts, while mostly intentional, also might have been inspired by a comment or question from a participant, and, yes, we did have examples of the various strategies that we often mention. At the same time, our editor, Lauren, to whom we owe so much thanks and gratitude, asked us if we would ever consider doing a book of consumables for educators to use. Hence, the seed was sown that finally grew into what became this book series.

Everyone acknowledges that communicating in mathematics is essential. Communication was one of the original five process standards of The National Council of Teachers of Mathematics. Over the years, we have collected, found, and created many classroom resources that provide authentic opportunities for students to communicate mathematically. So, the next step was culling through the many resources we had used and developed over the years. We sorted, resorted, looked at, accepted some, rejected others, and even created new ones as needed. We wanted to offer this series in four books to meet the individual needs of the various grade bands. At the same time, we wanted to provide examples and prompts that would cover the breadth of the grade

Afterword

band's mathematical topics and provide materials to support deepening the understanding of the topics.

Then to the writing and pulling together of the resources, which for us as mathematicians, the latter was much less challenging. This is ironic since this book focuses on writing and communicating in mathematics! What emerged is what you have on these pages. We hope that this resource becomes a go-to to meet your everyday classroom needs for providing opportunities for your students to engage in communicating about mathematics and communicating mathematically. Take what we have provided, expand on it, and make it your own. As you reimagine, retool, and even create your versions, we ask that you reach out and share. We would love to hear from you. You can find us at https://tljconsultinggroup.com/about-us/tammy-jones/ and https://leslietexasconsulting.com/about-leslie/.